Conservation and Biology of Small Populations

CONSERVATION AND BIOLOGY OF SMALL POPULATIONS

The Song Sparrows of Mandarte Island

Edited by

James N. M. Smith
Lukas F. Keller
Amy B. Marr
Peter Arcese

OXFORD
UNIVERSITY PRESS
2006

OXFORD
UNIVERSITY PRESS

Oxford University Press, Inc., publishes works that further
Oxford University's objective of excellence
in research, scholarship, and education.

Oxford   New York
Auckland   Cape Town   Dar es Salaam   Hong Kong   Karachi
Kuala Lumpur   Madrid   Melbourne   Mexico City   Nairobi
New Delhi   Shanghai   Taipei   Toronto

With offices in
Argentina   Austria   Brazil   Chile   Czech Republic   France   Greece
Guatemala   Hungary   Italy   Japan   Poland   Portugal   Singapore
South Korea   Switzerland   Thailand   Turkey   Ukraine   Vietnam

Published by Oxford University Press, Inc.
198 Madison Avenue, New York, New York 10016
www.oup.com

Oxford is a registered trademark of Oxford University Press

Library of Congress Cataloging-in-Publication Data
Conservation and biology of small populations: the song sparrows of
mandarte island / edited by James N.M. Smith . . . [et al.].
    p. cm.
Includes bibliographical references and index.
ISBN-13 978-0-19-515936-3
ISBN 0-19-515936-5
1. Song sparrow—British Columbia—Mandarte Island. 2. Song sparrow—
Conservation—British Columbia—Mandarte Island. 3. Bird populations—
British Columbia—Mandarte Island. I. Smith, James N. M., 1944–2005
QL696.P2438C66 2006
591.78'8—dc22   2005007142

9 8 7 6 5 4 3 2 1

Printed in the United States of America
on acid-free paper

We dedicate this book to the memory of Jamie (James N.M.) Smith, a warm friend, generous colleague and inspiring scholar. Jamie loved Mandarte and its song sparrows, and shared his enthusiasm for the island and its birds with us all. Sadly, although Jamie was the driving force behind this book, drew the sketches at the beginning of each chapter, and oversaw the final revisions, he did not live to see the manuscript published.

LFK, PA, ABM, JMR, WMH, & KO
July 2005

PREFACE

As the human impact on the earth's surface expands and intensifies, populations of many wild species are being squeezed into smaller and smaller patches of habitat. As a consequence, they face an increasing threat of extinction. As these threats are recognized, national and international conservation groups rush to add these populations, subspecies, and species to their "endangered" and "threatened" lists. In nations with strong conservation laws, listing often triggers elaborate plans to rescue these small and declining populations and restore their habitats. This book is about the demographic, ecological, and genetic factors affecting these populations, and the implications for population dynamics and persistence.

Theory suggests that small populations are liable to go extinct as a consequence of both demographic and genetic constraints. First, small size increases the likelihood that chance variation in births, deaths, and sex ratios will suddenly leave a population with no hope of survival. Second, small size is expected to foster chance loss of genetic variation, fixation of deleterious alleles, and reduced fitness when individuals are forced to mate with close relatives. At some level of reduced average fitness, a population will no longer be able to maintain itself. Third, extreme environmental events such as droughts and storms are more likely to wipe out a small population than a large one. Finally, these various mechanisms may interact with one another to increase the likelihood of population extinction.

In this book, we review these theoretical ideas, the existing empirical data, and explore the question, How well do small and isolated populations actually perform? We focus closely on a particular case, the song sparrows of Mandarte Island, British Columbia. This small population is particularly suitable for detailed study of the genetic and demographic threats to population survival. It is small enough and isolated enough that all individuals can be uniquely marked and their survival and reproduction can be monitored over many generations. Variation in nuclear DNA markers allows the tracking of genetic variability over time.

Our principal findings were that environmental catastrophes, not chance variation in reproduction and survival of individuals, were the main threats to the Mandarte sparrow population. In these catastrophes, extinction was narrowly avoided, rare genes were lost, and inbreeding depression in fitness occurred. However, average fitness remained high, and the depleted gene pool of the population was soon restored by immigration.

Many people helped to bring this book to completion. Charles Krebs and Dennis Chitty encouraged Jamie Smith to write a book in the first place. Our editors at Oxford University Press, Kirk Jensen, Peter Prescott, Anne Enenbach, and Kaity Cheng, assisted in many ways. Barbara Oberholzer spent many hours cross-checking references, creating the literature list, and preparing data for the figures. Pierre Brauchli and Jamie Smith expertly drew all the figures, and Judith Anderson prepared the index. André Dhondt and Amy Wilson read the whole manuscript and provided many useful suggestions. Jane Reid reviewed most of the chapters where she was not an author. Judy Myers, Simone Runyan, and Liana Zanette each commented helpfully on a draft chapter. Peter and Rosemary Grant and Nathaniel Wheelwright provided unpublished data on fluctuations of numbers of island birds. Charles Krebs suggested the use of generalized additive models, and Stephen Ewing gave SAS programming advice. Isla Myers-Smith helped with graphics and data management.

Roger Downie of the Department of Ecology and Evolutionary Biology, University of Glasgow, and David M. Bryant, formerly of the Department of Biology, University of Stirling, kindly provided space and facilities to Jamie during the preparation of the manuscript. Margery Smith, Liz Smith, and Bob Jackson also provided him with generous hospitality and emotional support in Scotland. James Berger and Bill Milsom of the Department of Zoology, University of British

Columbia, provided Jamie with vital relief from teaching. Judy Myers and Sarah Stead moved mountains to make this book a reality.

Frank Tompa, Rudi Drent, and Peter Grant encouraged Jamie to study the song sparrows of Mandarte Island and gave invaluable starting advice. Throughout the three decades of this project, our colleagues and students at the Universities of British Columbia, Wisconsin, and Glasgow and at the Zoology Museum in Zürich provided a stimulating academic environment. Colleagues and friends at these universities and elsewhere gave much encouragement, expert knowledge, and advice. Special mention here goes to Mark Beaumont, Mike Bruford, Dennis Chitty, Jim Crow, Carter Denniston, Peter Grant, Rosemary Grant, Dennis Heisey, Don Ludwig, Kathy Martin, Ken Petren, Dolph Schluter, Tony Sinclair, Steve Stearns, Donald Waller, and Michael Whitlock.

Our research on Mandarte Island was supported by grants and fellowships from the Natural Sciences and Engineering Research Council of Canada, the U.S. National Science Foundation, the Swiss National Science Foundation, the Roche Research Foundation, the University of Wisconsin, the Natural Environment Research Council (U.K.). The Tsawout and Tseycum First Nations gave permission to work on Mandarte Island.

A project of this magnitude and duration can be carried out only by generations of skilled and dedicated fieldworkers. We are deeply indebted to the following people for assistance with fieldwork on Mandarte Island and the surrounding islands: Pieter Bets, Ree Brennin (Helbig), Alice Cassidy, Kurt Cehak, Christine Chesson, Danielle Dagenais, Andrew Davis, Krista DeGroot, André Dhondt, Mike and Mary Dyer, Kathleen Fitzpatrick, Peter Grant, Sarah Groves, Lorne Gould, Dave Grosshuesch, Margret Hatch, Irene Heaven, Franziska Heinrich, Sara Hiebert, Sheila Hill, Wesley Hochachka, Marie-Josée Houde, Bruce Hutcheon, Kathryn Jeffery, Wendy Jess, Andrew Johnston, Gwen Jongejan, Durrell Kapan, Anne Labbé, Rob Landucci, Joanna Leary, Ken Lertzmann, Arnon Lotem, Marlene Machmer, Barbara Martinez, Juan Merkt, Bob Montgomerie, Arne Mooers, Joan Morgan, Richard Moses, Jean Munro, Judy Myers, David Newell, Kathleen O'Connor, Jaroslav and Anna Picman, Tarmo Poldmaa, Juanita Ptolemy (Russell), Cathy Redsell, Doug Reid, Jane Reid, Dick Repasky, Carol Ritland, Chris Rogers, Simone Runyan, Carolyn Saunders, Dolph Schluter, Doug Siegel-Causey, Allan Stewart, Philip Stoddard, Terry Sullivan, Mary Taitt, Anja Tompa, Harry van Oort, Nico Verbeek, Lotus Vermeer, Michaela Waterhouse, David Westcott, Amy Wilson, Scott Wilson, Yoram Yom-Tov, Liana Zanette, and Reto Zach. We also acknowledge any other helpers unwittingly omitted from this list.

This book is dedicated to the song sparrows of Mandarte Island. Watching these melodious singers gave us countless hours of pleasure.

JNMS, LFK, ABM, PA
March 2005

CONTENTS

CONTRIBUTORS

Peter Arcese
Department of Forest Sciences and Centre for Applied
Conservation Research
University of British Columbia
#3041 - 2424 Main Mall
Vancouver, British Columbia
V6T 1Z4, Canada
E-mail: peter.arcese@ubc.ca

Wesley M. Hochachka
Bird Population Studies Program
Laboratory of Ornithology
Cornell University
159 Sapsucker Woods Road
Ithaca, New York 14850-1999
USA
E-mail: wmh6@cornell.edu

Lukas F. Keller
Zoologisches Museum
Universität Zürich
Winterthurerstr. 190
CH-8057 Zürich
Switzerland
E-mail: lfkeller@zoolmus.unizh.ch

Amy B. Marr
Department of Forest Sciences and Centre for Applied
Conservation Research
University of British Columbia
#3041 - 2424 Main Mall
Vancouver, British Columbia
V6T 1Z4, Canada
E-mail: amymarr@alumni.princeton.edu

Kathleen O'Connor
Department of Forest Sciences and Centre for Applied
Conservation Research
University of British Columbia
#3041 - 2424 Main Mall
Vancouver, British Columbia
V6T 1Z4, Canada
E-mail: kathleen_oc@yahoo.com

Jane M. Reid
Jesus College and Department of Zoology
University of Cambridge
Downing Street
Cambridge, CB2 3EJ
United Kingdom
E-mail: j.reid@jesus.cam.ac.uk

James N. M. Smith (Deceased)
Dept. of Zoology and Beaty Biodiversity Centre
6270 University Blvd.
University of British Columbia
Vancouver, British Columbia
Canada, V6T 1Z4

Conservation and Biology of Small Populations

1   Genetics and Demography of Small Populations

James N. M. Smith and Lukas F. Keller

In the above sketch, a male song sparrow sings on a tiny rocky islet in the Gulf Islands Archipelago. His mate forages in the foreground. No other song sparrows inhabit the islet. Populations of only one male and one female are an extreme case for a sexually reproducing species, but conservation biologists often have to manage populations little bigger than this. This book is about the demography and genetics of such small populations.

Populations are collections of individuals of the same species that occupy a defined place (Krebs 2001). In sexually reproducing species, they can vary in size from a single pair of breeders on an island to billions of migratory locusts. Some populations, like locusts, are highly mobile and extend over vast areas. Others live discontinuously in patchy habitats, on mountaintops, or on island archipelagoes. Collections of subpopulations are sometimes termed *metapopulations* (Hanski 1999), where movements of individuals link the dynamics of adjacent subpopulations.

All populations fluctuate, growing when reproduction and immigration exceed mortality and emigration, and declining when the reverse conditions apply. When numbers fluctuate in a large population, local extinction (hereafter *extirpation*) is unlikely, because it is likely that some individuals will survive even the harshest conditions. However, fluctuations can cause small populations to disappear or decline to a size where genetic decay or random loss become likely. If only one population of the species remains, global extinction can follow. The behavior of small and isolated populations is thus of central interest to conservation biologists whose aim is to save rare and declining populations. In this book, we examine how small size influences the genetic structure and ecological performance of populations. In particular, we use a 28-year study of a small songbird, the song sparrow, on a tiny island in western Canada to illustrate the effects of chance processes on the ecology and genetic composition of the population.

Small Populations and the Extinction Crisis

The activities of modern humans are causing the loss of many populations and species (e.g., Stearns and Stearns 1999). As human populations continue to grow (Cohen 2003), we are usurping the planet's production to maintain our increasing numbers. The bare facts are now widely evident (e.g., Wright 2004). Overharvesting has caused the decline of many populations (Caughley and Gunn 1996), and continues to threaten many marine (e.g., Hobday et al. 2001) and terrestrial (e.g., Fa et al. 2004) organisms. Fragmentation and loss of habitats are reducing once large populations to smaller and more scattered ones (Schroeder et al. 1999; Gaston et al. 2003). Recent human-induced climate change has already extirpated some populations and is threatening many others (Thomas et al. 2004). In sum, these processes amount to a deepening global extinction crisis and a consequent need to manage increasing numbers of small and declining populations.

Bird Populations on Islands and Conservation

Island birds have been prominent targets in the conservation of threatened and endangered populations for several reasons. Most of the currently endangered bird species occur on islands (BirdLife International 2000; Bell and Merton 2002), and most historical and recent prehistorical avian extinctions have occurred on islands (Steadman 1995; Caughley and Gunn 1996). Birds on islands exhibit varied and sometimes surprising demographic patterns. In some cases, they have shown rapid recovery in numbers from catastrophic losses in storms (McCallum et al. 2000). In other situations, fluctuating environmental conditions have led to unpredictable long-term changes in population size (Grant et al. 2000). If individuals cannot disperse readily from islands, they may be forced to breed with close relatives. Thus, detrimental consequences of inbreeding can be measured in island populations (Keller and Waller 2002). Finally, islands are ideal for studying the establishment and growth of translocated groups of birds (Komdeur 1997; Armstrong and Ewen 2002) and for attempting extreme management options, such as predator eradication (Myers et al. 2000). In the following five sections, we discuss the principal theoretical ideas pertaining to the genetics and demography of small populations.

1.1. Demographic Risks to Small Populations

One factor that puts small populations at particular risk of extirpation is *demographic stochasticity* (i.e., variability), or chance effects on births and deaths of individuals. The dusky seaside sparrow is a vivid example. This critically rare subspecies of songbird occurred in salt marsh habitats near Cape Kennedy Space Station, Florida. By 1979, sparrow numbers had fallen so low that managers decided to capture all wild individuals for a captive breeding program. Five of the last six birds were captured successfully, but all five proved to be males, thus dooming the subspecies to extinction (Post and Greenlaw 1994).

In the above example, chance variation in sex ratio is sufficient to explain population loss, but is there broader evidence linking small size with rapid extirpation? In fact, few quantitative studies have demonstrated this relationship (however, see chapter 11). One much-quoted example describes the persistence of 355 populations of land birds inhabiting 16 small offshore islands near Great Britain over a 30-year period. Extirpations occurred frequently, and the rate of extirpation increased with variation in size of the populations (Diamond 1984; Pimm et al. 1988). Most of the populations that were lost consisted of less than 10 breeding pairs. A 50-year study in the western United States examined size and extirpation rate in 122 bighorn sheep populations. Berger (1990) found that all populations that started with fewer than 50 individuals became extirpated, whereas populations starting with more than 100 individuals nearly all persisted. The bighorn sheep case seems a perfect illustration of a "small is vulnerable" rule, with an extirpation threshold of 50 individuals. However, it is uncertain whether chance demographic variability alone caused extirpation of the smaller populations or whether some other cause, e.g. here respiratory diseases acquired from domestic sheep, did so (Monello et al. 2001; Cassirer 2005).

In summary, very small populations are vulnerable to chance extirpation from random effects. It is less clear, however, how small populations must become before chance extirpation becomes likely. Morris and Doak (2002) suggest that populations below 20 individuals are particularly vulnerable, and Berger's (1990) study hints that 50 individuals might represent an upper bound of extirpation risk from random causes.

1.2. Environmental Risks to Populations

Early population biologists such as David Lack (1954, 1966) believed that, while populations are influenced by environmental perturbations, regulatory influences tend to keep numbers near an equilibrium carrying capacity. Others argued, however, that environmental variation has an overwhelming effect on population size (e.g., Andrewartha and Birch 1954). It is now widely accepted that both regulatory processes and environmental perturbations affect numbers (e.g., McCallum et al. 2000; Clutton-Brock and Pemberton 2004), although some species and populations are more subject to environmental influences than others (Lande et al. 2003).

Environmental stochasticity in weather has marked effects on population size (Lande et al. 2003). Hurricanes strike Heron Island, a coral cay on the Australian Great Barrier Reef, in late summer about every 8 years, killing up to 50% of the 400 or so Capricorn silvereyes on the island (McCallum et al. 2000). The El Niño Southern Oscillation causes catastrophic mortality and reproductive failure in breeding seabirds in the western Pacific (Schreiber and Schreiber 1989) and also promotes explosive population growth in land birds in the eastern Pacific (Grant et al. 2000). However, even extreme weather events rarely cause medium or large-sized populations to approach extirpation.

Epidemic diseases also affect animal numbers (Hudson et al. 2002) and may have caused extinctions in Hawaiian honeycreepers (van Riper et al. 1986; Cann and Douglas 1999) and Australian and Panamanian rainforest frogs (Laurance et al. 1996; Berger et al. 1998). A disease did cause the extinction of a captive population of land snail *Partula turgida* (Cunningham and Daszak 1998). It is, however, often difficult to be sure that parasites or diseases are a principal cause of population losses (Freed 1999; Sures and Knopf 2004) rather than a symptom of other stresses.

Another strong environmental effect on island birds is the arrival of novel predators (e.g., Blackburn et al. 2004). Introduced brown tree snakes extirpated most of the endemic birds of the Pacific island of Guam between 1960 and 1985 (Savidge 1987). Introduced competitors on islands can also threaten populations (Petren and Case 1996), but few examples have been documented.

In summary, environmental variability influences the size of both small and large populations and may cause local extirpations of small populations (Morris and Doak 2002; Lande et al. 2003).

1.3. Allee Effects

An "Allee effect" describes a situation where population growth rate declines progressively as population size declines (Allee 1931; Engen et al. 2003). Allee effects can be caused by difficulties in finding mates when populations are sparse, or by groups falling below a threshold size needed to reproduce successfully or gain protection from predators. For example, some seabirds (e.g., northern gannets, Nelson 1978) reproduce better in large colonies than in small ones. Cooperative social systems in mammals may break down in small groups, such that mortality exceeds recruitment (Rasa 1989; Courchamp and Macdonald 2001).

While Allee effects fascinate theoreticians (e.g., Liebhold and Bascompte 2003), evidence that they occur in wild populations comes mainly from highly social animals (see above) and plants that depend on rare animals for pollination (e.g., Groom 1998). However, the nonsocial Granville fritillary butterfly also exhibits Allee effects. Small fritillary populations on islands have high emigration rates and low mating success in females, and therefore negative population growth rates (Kuussaari et al. 1998). It remains uncertain how much Allee effects influence small populations in the wild. Compelling evidence that they commonly do so is limited, although some have argued otherwise (Stephens and Sutherland 1999).

1.4. Genetic Risks to Small Populations

Small populations also face genetic risks that often arise during or just after sudden reductions in population size, or *population bottlenecks*. Bottlenecks caused by natural catastrophes or human impacts are common among island bird populations (Briskie and Mackintosh 2004). In an extreme example, the Mauritius kestrel was reduced to four individuals and only two breeders by 1974. The original population size of this small falcon is unknown, but its distribution is thought to have covered the entire 2,040 km² island (Groombridge et al. 2001). Although the population has since recovered to several hundred individuals (Jones et al. 1995), it has not regained the genetic variation lost in the past bottleneck (Groombridge et al. 2001; S. Ewing, personal communication). Thus, genetic diversity in isolated populations is not only a function of their current size but also a legacy of past bottlenecks in numbers (Frankham et al. 2002).

When we census individuals in a population, we do not expect all individuals to contribute genetically to the next generation because some fail to mate or produce offspring. It is therefore useful to define the *genetically effective population size* (N_e), the number of individuals that actually breed and contribute to future generations. Effective population size is generally smaller than actual population size and can be up to 10 times smaller in polygamous mating systems (Frankham et al. 2002; see also chapter 6). Therefore, genetic risks can affect populations that face little risk of extirpation from demographic stochasticity.

Small populations face four main genetic risks. First, when a larger population first becomes small, alleles will be lost randomly and others may become fixed via *genetic drift* (Wright 1977). If the population remains small, further alleles may be lost through drift. Reduced quantitative genetic variation may reduce the ability of small populations to adapt to changing conditions. For example, forcing laboratory populations of fruit flies through bottlenecks reduces their ability to adapt to salt stress (Frankham et al. 1999).

There is a solution to the effects of drift in small populations. New individuals can be introduced to the target population to restore the lost genetic variability (e.g., Madsen et al. 1999; Vilà et al. 2003; see chapter 8). However, there is a second genetic risk associated with such introductions. If the introduced individuals are too different genetically, the hybrid offspring of local individuals and the introduced genotypes may exhibit *outbreeding depression*, that is, lowered fitness (Templeton 1986). In extreme cases, population extirpation may occur (Hughes et al. 2003).

Third, in small populations random events will have stronger effects than selection on the genetic composition of future populations. Thus, mildly to moderately deleterious mutations will tend to accumulate and reduce the mean population fitness. This, in turn, could reduce population size and increase the accumulation of harmful mutations. In the short term, theoretical analyses suggest that populations with N_e of less than 100 could become extirpated from this *mutational meltdown* (Lynch et al. 1995). In the long run, even populations with N_e of a few thousands might be endangered by this mechanism (Lande 1998). Likely examples of mutational meltdown in wild populations have recently begun to accumulate (e.g., Rowe and Beebee 2003; Puurtinen et al. 2004).

The loss of genetic variation and the accumulation of mutations are likely to reduce fitness gradually over many generations. However, populations face a final, and more immediate, genetic threat. Small population size restricts opportunities for mating and promotes inbreeding among relatives. Inbreeding increases the frequency of individuals that are homozygous for alleles that are identical by descent (Wright 1977). Some of these alleles will be deleterious and their effects will be expressed when two copies are present at the same locus. If inbred progeny have reduced fitness (i.e., *inbreeding depression*), population viability can fall. For example, inbreeding depression in the Glanville fritillary butterfly is expressed as reduced egg hatchability. When fewer eggs hatch per clutch, larval groups are smaller, and larval survival is reduced (Nieminen et al. 2001). Thus, increasing inbreeding levels make fritillary populations more likely to disappear, even after controlling for the effects of population size (Saccheri et al. 1998). Although any of these four genetic mechanisms could threaten population persistence, inbreeding poses the most immediate risk (Keller and Waller 2002).

How small do threatened populations need to be before they face extirpation from inbreeding? Early conservation biologists proposed rules-of-thumb to answer this question. For inbreeding depression, Franklin (1980) and Soulé (1980) recommended a minimum N_e of 50. This value was based on breeding experiments in chickens and pigs and on theoretical studies of the balance between the costs of inbreeding in a trait and selection to purge the deleterious alleles involved. These studies suggested that, when inbreeding levels increase by more than about 1% per generation (which is likely to happen when $N_e < 50$), purging selection is too weak to compensate for the negative effects of inbreeding.

Therefore, to assess the genetic risk from inbreeding we need to know the magnitude of inbreeding depression in a trait, which can be estimated by comparing the fitness of inbred and outbred individuals. We also need to know the answer to a more difficult question: Can the detrimental alleles causing inbreeding depression be purged by selection before the population is extirpated? We explore these issues in chapters 7 and 11.

1.5. Integrating Demographic and Genetic Influences on Small Populations

The Extinction Vortex

In 1986, Michael Gilpin and Michael Soulé coined an enduring metaphor, the *extinction vortex*, to describe synergistic interactions that might cause extirpation of declining populations. The extinction vortex proposes that a large population first becomes reduced in size by environmental catastrophes, overharvesting, or habitat loss. At its new smaller size, the population becomes vulnerable to inbreeding and further environmental variability, thus lowering numbers again (figure 1.1, left). When the population becomes very small, demographic stochasticity, further inbreeding depression, mutation accumulation, or Allee effects (Lande 1998) drive it to extirpation.

No population has undergone the extinction vortex while being studied in detail. There is, however, a suggestive early case. The heath hen was a small grouse that was once distributed from Maine to Virginia in the eastern United States. By 1870, it had been extirpated from the mainland and occurred only on the island of Martha's Vineyard in Massachusetts. In 1907, a habitat reserve was created on the island and the population recovered rapidly from 50 to 2,000 by 1915. However, a wildfire that year, followed by harsh winter weather, an invasion of northern goshawks, and a poultry disease reduced numbers precipitously. By 1927, the 13 survivors were mostly males, and the population disappeared by 1932 (Bent 1932; Simberloff 1988). This example illustrates a key property of extinction vortices; extinction occurred,

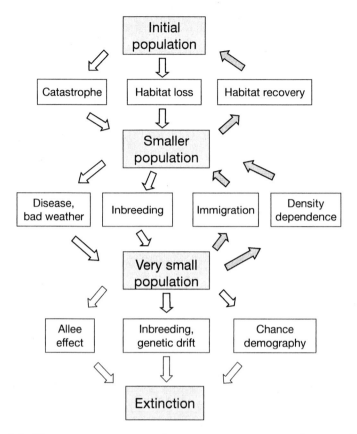

Figure 1.1. The extinction vortex. Open arrows represent influences that depress population size. Shaded arrows represent influences promoting population recovery.

even though the cause of the initial decline, habitat loss, was addressed successfully. Simberloff (1988) cites other suggestive cases.

The extinction vortex is often used to illustrate runaway biodiversity loss (e.g., Guerrant 1992). However, when populations become small, they generally recover (figure 1.1, right) because of negative density dependence, reduced harvesting, or immigration. Several bird examples are reviewed by Newton (1998, pp. 477–479), but perhaps the most spectacular case of recovery is for a marine mammal. Northern elephant seals in the eastern Pacific were reduced to between 8 and 20 individuals between 1884 and 1892. However, by 1991 they had recovered to 127,000 individuals, and numbers were still increasing at 6% annually (Stewart et al. 1994).

Population Viability Analysis

In principle, the combined effects of demographic and genetic threats to populations can be studied by integrated modeling. Indeed, modeling has always played a central role in both population biology (Levins 1966; Lande et al. 2003) and population genetics (Fisher 1930; Wright 1969). Early population models were deterministic and often based on equilibrium assumptions because of simplicity and mathematical tractability. However, although chance processes were well known to influence both evolution and ecology (Wright 1977; Levins 1969), the tools needed to incorporate chance variation into population models were lacking before 1970. Then, the computer age facilitated the development and use of stochastic population models, the field now known as population viability analysis.

Population viability analysis (PVA) is a set of modeling procedures that are used widely in the conservation and management of populations (Beissinger and McCullough 2002). PVA models simulate the performance of populations using data on survival and reproduction and estimates of the variability in these traits. Beissinger and McCullough (2002), Morris and Doak (2002), and Lande at al. (2003) provide detailed recent surveys of the principal techniques used in PVAs. A brief history of the development of these techniques follows.

The practice of modeling how demographic variability is likely to affect the extirpation of populations was begun by MacArthur and Wilson (1967), and developed further by Richter-Dyn and Goel (1972), Shaffer (1983), and Goodman (1987). How environmental variation affects extirpation was first modeled by Reddingius and Den Boer (1970), followed by (Leigh 1975, 1981) and Strebel (1985). Demographic and environmental variability were then modeled together by Shaffer and Sampson (1985) and Goodman (1987).

PVA modeling has been used to show that both demographic and environmental influences increase the variance in population growth rates, particularly in density-independent models. This increased variance can cause extirpation when variability is high and when populations are isolated from immigration. The extirpation rates predicted in PVA models have influenced the management of some high-profile endangered species, notably the northern spotted owl in the United States (Lande 1988b; Franklin et al. 2000).

The use of PVA models to predict the future performance of populations, however, is controversial. For example, Donald Ludwig (1999) considered that predictions of time to extirpation are dubious, particularly when they are based on short runs of population data. Later, he and others noted that vital information is usually lacking on how catastrophic events influence populations (Ellner et al. 2002). A further criticism is that many PVA models lack biological realism. In one such example, a model that predicted rapid population loss in Capricorn silvereyes (Brook and Kikkawa 1998) was superseded by a more realistic model, which used the same data to predict that the population would persist almost indefinitely (McCallum et al. 2000). Finally, Ludwig and Walters (2002) claim that PVA models are also far from being able to connect population dynamics meaningfully to policy options.

However, new methods in PVA are being developed (e.g., Engen and Sæther 2000) and applied (e.g., Sæther et al. 2002). PVA modeling is increasingly being used to ask which factors within a manager's control should be targeted to enhance a population's survival (e.g., Lindenmayer et al. 1993; Reed et al. 2002), rather than to predict a time to, or a probability of, extirpation. PVA models also offer the promise of using demographic and genetic data jointly to assess future population performance. This promise, however, has yet to be widely realized. One reason for this failure is that the necessary genetic data are usually missing for species of conservation interest (Allendorf and Ryman 2002).

Which Matters Most—Genetic or Demographic Risks?

Early in the development of conservation biology, genetic risks to small and captive populations received close attention (e.g., Franklin 1980; reviewed in Simberloff 1988). However, Russell Lande (1988a) wrote a provocative and influential paper on the northern spotted owl that attacked the dominance of genetic thinking in conservation biology. Lande argued that demographic and environmental stochasticity were far more important and immediate causes of population extirpation than genetic deterioration.

A few years later, Graeme Caughley (1994) expanded on Lande's thesis in an even more provocative manner. Caughley stressed a need for strong empirical studies of the causes of population declines, rather than theoretical estimates of extirpation risk for small populations. Building from this starting point, he favored the second of two paradigms of conservation biology, the *small population paradigm* and the *declining population paradigm*. By the former, he meant the genetic and demographic theory of extirpation at small population size that we have just reviewed. In contrast, the declining population paradigm was a hypothesis-driven search for the best mechanistic explanations for population declines. By distinguishing these two paradigms, Caughley laid bare an underlying tension between modelers and field biologists interested in conserving populations. Caughley's paper also aroused the ire of conservation biologists with a genetic bent (e.g., Hedrick et al. 1996). In chapter 11, we reassess the claims that genetic considerations matter less than demographic ones when assessing extirpation risk in small populations.

The Scope of This Book

Since Caughley's paper in 1994, there has been an explosive growth in knowledge about the biology of small populations. We now know that strong inbreeding effects in nature are relatively common (Keller and Waller 2002). There have been detailed and long-lasting studies of the performance of small populations of butterflies (Ehrlich and Hanski 2004), birds (e.g., McCallum et al. 2000; Sæther et al. 2002; Keller et al. 2002), mammals (Clutton-Brock and Coulson 2002; Creel and Creel 2002), and reptiles (Madsen et al. 1999).

We are therefore in much a better position than in 1994 to assess the joint influences of demographic and genetic mechanisms on populations. We present here an integrated account of the effects of genetics, environment, and demography on a focal small population, the song sparrows of Mandarte Island, British Columbia, Canada. We do not claim that the results from the sparrows of Mandarte Island, or indeed, any other population, can be generalized uncritically (see chapter 11). We do believe, however, that detailed case studies of genetics and demography of small populations are vital to developing a robust theory of extinction. In particular, detailed studies of inbreeding, environmental stochasticity, and immigration can provide insights that more broad-brush studies cannot.

In chapter 2, we describe our study species and principal study site. The life history of the song sparrow, which it shares with many other open-cup nesting songbirds, is described in chapter 3. In chapter 4, we describe and analyze the marked fluctuations in numbers shown by song sparrows over our 28-year study. These fluctuations revealed frequent catastrophic losses, varying but repeating demographic patterns, and also long-term unpredictability. In chapter 5, we consider how an invasive alien species, the brood parasitic brown-headed cowbird, affected the song sparrow on Mandarte Island. Chapter 6 deals with the social behavior and mating system of the song sparrow, key attributes for understanding both genetic and demographic patterns. In 1987, we began to collect blood samples from the population, just before a severe bottleneck in numbers occurred. Assays of marker genes using these samples allowed us to document the loss of genetic variation in the 1989 bottleneck, its subsequent recovery, and the spread of immigrant genes through the population (see chapters 7, 8). In chapter 9, we integrate several life history and demographic issues raised in preceding chapters by considering the lifetime reproductive success of individuals. In chapter 10 we use a population viability model to consider how the persistence of the Mandarte population depends on immigration, catastrophic mortality, the invasion of cowbirds, and chance variation in reproduction and juvenile survival. Finally, chapter 11 summarizes our findings and discusses their implications.

2 Song Sparrows, Mandarte Island, and Methods

James N. M. Smith

Mandarte Island is a six-hectare islet in the San Juan/Gulf Islands archipelago on the western Canada/U.S. border. Since 1959, and continuously since 1974, four generations of researchers have gone there to study the population biology and genetics of a small population of song sparrows.

2.1. The Study Species

The song sparrow, *Melospiza melodia*, belongs to the family Ember-izidae, which includes the buntings and North American sparrows (Byers et al. 1995; Rising 1996). It breeds over most of North America south of the tree line, and it is often locally abundant.

Song sparrows (figure 2.1b) have a brown back, rounded tail, gray and brown striped crown and face, and a brown-streaked breast with a bolder central spot. Males and females are similar in plumage, but adult males (26 g) average about 8% heavier and have a 4% longer wing chord (69 mm) than do adult females on Mandarte Island. Older birds have slightly longer wings than first-year birds but fractionally shorter legs (Smith et al. 1986). Song sparrows vary considerably in size across their range (Aldrich 1984). Males from the Aleutian Islands weigh up to 46 g while male *Melospiza melodia samuelis* from San Francisco Bay weigh only 18 g (Arcese et al. 2002; Chan and Arcese 2003). Plumage also exhibits extensive geographic variation in song sparrows (Arcese et al. 2002) and is used to classify subspecies, some of which are of conservation interest. This plumage variation, however, does not map closely with geographic variation in genetic markers (Fry and Zink 1998; Chan and Arcese 2002).

Despite its modest visual credentials, people notice and value the song sparrow because of its distinctive and conspicuous behaviors: an arresting four-part song, pumping of the tail in flight, a loud *tchunk* alarm call, pugnacity toward other songbird species (Nice 1943), and a boldness when humans are present. These behavioral traits allow even novice birders to distinguish the song sparrow from other similar-looking sparrows.

Song sparrows are generally migratory in the northern half of their range, but most western coastal populations are resident (Arcese et al. 2002). The Mandarte population is largely resident, but up to 20 migrant individuals identifiable by their pale-colored plumage also visit Mandarte in winter. It is also possible that a few Mandarte breeders migrate.

The song sparrow became one of the world's best-known songbirds in the 1930s, when population biology was in its infancy. Margaret Morse Nice (1937, 1943) used colored leg bands to mark all birds individually. She then followed the behavior, numbers, and reproduction of a population in suburban Columbus, Ohio, closely for 7 years. Modern behavioral and population ecologists still employ her approach.

Among Nice's many novel findings about the song sparrow were that brown-headed cowbirds and territoriality influence sparrow numbers. We revisit these topics in chapters 5 and 6.

Because Nice was a (very professional) "amateur," and because of World War II, her groundbreaking work was not immediately followed up. However, work on song sparrows resumed after the war. Marshall (1948) and Johnston (1956a, 1956b) studied populations in San Francisco Bay, and Tompa (1964) worked on Mandarte Island. Beginning in the 1960s, the song sparrow became a popular and enduring subject for studies of song development. In the 1980s, it also became a model organism for studying the hormonal basis of territoriality (reviewed in Arcese et al. 2002).

The song sparrow breeds in suburban gardens (Nice 1937), old fields (Robinson et al. 2000), desert oases and watercourse margins (Patten et al. 2004), coastal dunes (Hare and Shields 1992), and salt marshes (Marshall 1948). In most of its breeding range, however, it is associated with open brushy habitats near water. In winter, migrant individuals use a wider range of habitats. Population density of the song sparrow varies greatly: Dense populations occur in salt marshes, clear cuts in the U.S. Midwest, and the shrub-dominated Mandarte Island. Densities are lower in most other habitats (Arcese et al. 2002).

Song sparrows, like most of the 4,500 species of songbirds worldwide (Sibley and Ahlquist 1990), build open-cup nests and commonly rear two or three broods per year (see chapter 3). Song sparrows are a "weedy" species and colonize moist early successional habitats readily as juveniles. They are generally sedentary as adults, although Weatherhead and Boak (1986) found low fidelity to territories and mates in a population in southern Ontario.

In summary, there is a wealth of comparative material on the song sparrows' biology (reviewed in Arcese et al. 2002). It has already served as a "model organism" for the study of population dynamics, singing, and territorial behavior and has potential to yield additional insights into conservation genetics (see chapters 7, 11).

2.2. The Study Area

Mandarte Island (48° 38′ N, 123° 17′ W) lies in the Gulf Islands (San Juan) Archipelago (figure 2.2). It is familiar to tourists who ride the Washington State Ferry from Friday Harbor, Washington, to Sidney, British Columbia, as "the bare rock with all the outhouses" (actually

(a)

(b)

(c)

Figure 2.1. (a) Aerial photograph of Mandarte Island taken in August 1975. Note the longitudinal band of shrubs, the trails cut through the shrubs to allow us to move around the island, and the camp buildings on the lower left shore. (J. Smith.) (b) Song sparrow on Mandarte Island. (L. Keller.) (c) Photograph of the sea island scrub that covers about one-third of Mandarte Island. Note thickets of blackberry, a fairly recent invader to the island, in the foreground. (L. Keller.) (d, e) Losses of mature trees between 1963 (P. R. Grant) and 2004 (P. Arcese) (f) The mosaic of sea island scrub, open meadows, and a few trees on Mandarte Island. Note the Nootka rose (right), Saskatoon (center back), willow (left), and grasses. (g) Observer looking for a song sparrow nest from a fruit-picking ladder in 1987. (J. Smith.) (h) Song sparrow nest containing a song sparrow and a cowbird egg. The cowbird egg is at the bottom. (L. Keller.) (i) Song sparrow and cowbird nestlings at about 6 days of age. (J. Smith.)

(d)

(e)

(f)

(g)

(h)

(i)

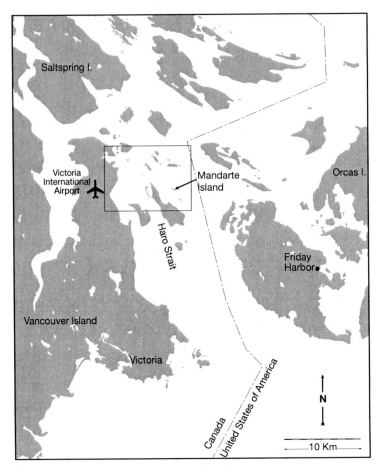

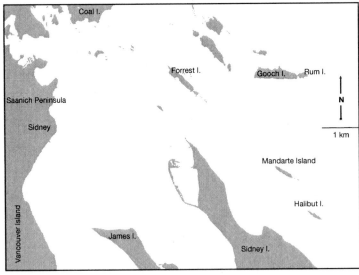

Figure 2.2. Maps of the Southern Gulf Islands and the islands close to Mandarte Island. The dashed line marks the Canada/U.S. border.

←───

observation blinds). It is approximately 700 m long and averages 85 m wide (figure 2.1a). Mandarte is part of the traditional territory of the Tsawout and Tseycum native peoples, who once foraged on the island for camas lily bulbs, seabird eggs, fish, and shellfish. Researchers have visited it for nearly 90 years to study the island's seabirds and song sparrows (Drent et al. 1964). A rustic field station was established on the island's north shore in 1957 (figure 2.1a). Our field team lives at the station for 3–4 months each summer, and we make short visits each fall and spring.

Mandarte lies about 25 km north of the city of Victoria on the west side of Haro Strait near the Canada/U.S. border. Several other low-lying islands, all of which support song sparrow populations, are found within 5 km (figure 2.2). The island is composed of sandstone and conglomerate rock, and soils are shallow or absent over much of the island. Like most neighboring islands, Mandarte tilts to the northeast, so the southwest side of the island is a 10–30 m cliff while the north side rises only 2–7 m above the high tide line. This orientation shelters most of the island from the prevailing southwest winds but exposes it to storms from the north and northeast.

Vegetation

Small Gulf Islands like Mandarte (~5 ha) support three main habitats for song sparrows: patches of dry forest and sea island scrub, open meadows, and cliffs. In forest patches, Douglas fir dominates, with some arbutus (madrone), Garry oak, and grand fir. Habitats on Mandarte consist of grassland (57%), open cliffs (21%), and sea island scrub and small trees (22%). The 1- to 3-m-high sea island scrub extends the whole length of the island (figure 2.1b,c). It is the principal breeding habitat of the song sparrow, and dominant shrubs include snowberry, Nootka rose, and Himalayan blackberry. Other shrubs and small trees in the scrub include Saskatoon, Garry oak, ocean spray, red elderberry, cut-leaf blackberry, chokecherry, and willow. The island is fertilized abundantly by thousands of seabirds, and the edges of the shrubs support a rich band of nettles, fireweed, fringecups, camas, and other forbs.

The lush early spring growth of grasses and forbs provides nesting cover and foraging habitat for the sparrows. The grassland, cliffs, and a fourth habitat, the rocky intertidal zone, provide foraging areas. Taller shrubs and trees provide perches for birds to sing from and survey their territories. However, the lush and moist spring vegetation typically dies back by July each year, reducing cover for molting sparrows. Grasses and forbs resume growing each October when the summer dry spell ends.

Vegetation Dynamics

The island's vegetation changed markedly during our study. The most obvious change was the loss of larger trees (figure 2.1d,e). All large live trees in 1963 died under a rain of guano from perching seabirds, and most have since fallen. Many of the small bitter cherries, which once provided a scattered emergent canopy 3–6 m in height over much of the island, have also fallen. These losses have reduced the number of perches available to singing male sparrows, perhaps affected the locations of territory boundaries and altered the searching behavior of parasitic cowbirds (see chapter 5).

The scrub margins have changed slightly, with considerable recruitment of patches of Himalayan blackberry (figure 2.1c). Blackberry thickets now make up about 10% of the sea island shrub. These thickets die back in cold winters and are replaced temporarily by tall forbs, particularly nettles. Ivy now dominates groundcover around the west-central patch of tall fir snags and willows. Spring grazing by Canada geese, which colonized the island in the 1980s, has reduced grass cover.

Climate

The study region lies in the rain shadow of the Olympic Mountains. It has mild wet winters and cool summers. Both precipitation (figure 2.3a) and temperature (figure 2.3b) are strongly seasonal. November to January are predictably wet, with monthly precipitation averaging 400 mm, and July to September are dry, with an average of 110 mm per month. Summer rains usually fall on only a few days per month, and some summer months have no precipitation at all. December and January are the coldest months with mean temperatures of 4.8°C and August is the warmest at 15.9°C. Extreme cold and heavy snowfalls occur only rarely, and snow cover generally melts quickly. However, extreme winter weather and climate warming can both influence song sparrows strongly (Tompa 1971; see chapters 3 and 4).

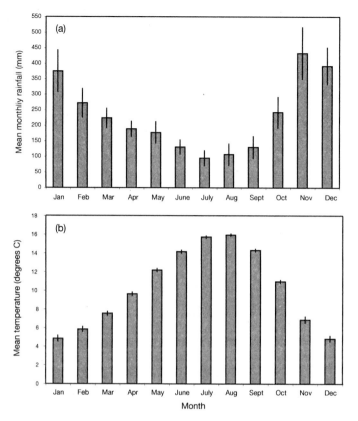

Figure 2.3. Mean monthly precipitation (a) and temperature (b) over 28 years (1974–2001) at Olga, Washington State, 35 km from Mandarte Island and at the same elevation. Error bars are 95% confidence intervals.

2.3. Methods

Jamie Smith asked Frank Tompa in 1974 if it would be possible to locate all the successful sparrow nests on Mandarte each year, Frank emphatically replied no and cautioned Jamie that the species was quite sensitive to disturbance when nesting. Nevertheless, Jamie arrived on Mandarte later that year to check out Frank's assessment, armed with the unworthy determination of the new Ph.D. to prove that older and wiser heads were wrong. At first, things did not go well; an exploding camp stove, several days of cold food and bad weather, and a pulse of nest failures lowered Jamie's morale and nearly convinced him that Frank was right. However, we eventually found that Mandarte song sparrows were no more sensitive to disturbance than the song sparrows previously studied successfully by Nice, Marshall, and Johnston (described above). We therefore adapted the following approach from Nice (1937) and used it in our subsequent work.

Our fieldwork began each year with surveys of banded birds in March and April. At this time, we mapped the locations of all territorial individuals and of any nonbreeding floaters present in the population. Once breeding began in late March or early April we monitored the breeding activities of each female bird weekly. Finding nests taxed our small research team fully in high-density years, but population crashes allowed us to study nearby islands. We also worked intensively in some winters to measure recruitment to territories, body condition, and survival.

Trapping and Banding

We first caught adults in mist nests in the fall and winter of 1974–1975. Thereafter, we banded most individuals as nestlings at 5–7 days of age with a numbered metal band and two to three plastic color bands. Some netting was required each year to capture immigrants and recapture other birds. However, more than 97% of recruits to the breeding population were color marked and of known parentage each breeding season simply because of the successful banding of nestlings.

Nest Finding

Most song sparrow nests are very well concealed by cover and can be almost invisible, even when one looks directly at them. However, after much trial and error we found a systematic way to overcome this difficulty. We observed females from above from rocky vantage points and fruit-picking ladders (figure 2.1g). Some females built nests in full view of observers. Others gave caterwaul calls when leaving the nest (Nice 1943; Arcese et al. 2002). Males gave us some clues to nest locations via loud songs given near the nest and pounces on females as they left the nest area. Song sparrow females incubate and feed on a predictable cycle (25–40 minutes of incubation, 5–10 minutes of foraging). This fact allowed one observer to search for two or more nests at once. Finally, the same general nest locations were reused across years, giving us prior information on where to look.

Although these methods worked well, the topography was unfavorable for nest finding in some spots. Other females were secretive and approached and left their nests quietly under cover. Thus, we did not find 28 successful nests over the 28 study years. When this happened, we trapped the resulting young in mist nests when they became mobile at about 20 days of age. In 1992, three unbanded fledglings could not be caught and remained of unknown parentage and geographic origin; fortunately, none of these birds became breeders. We also did not find some nests that failed early in the nesting cycle, causing us to underestimate the number of nesting attempts per female slightly.

Nest Checking and Estimating Fledging Success

Once nests were located, we checked them every 3–7 days, noting clutch sizes and the eggs of the parasitic brown-headed cowbird. While the two species lay similar eggs (figure 2.1h), sparrow eggs are typically more elongate, unevenly spotted with brown, and have a pale green ground. Typical cowbird eggs are more rounded at the blunt end, more pointed at the sharp end, and have smaller and more even reddish-brown spots on a cream ground (Smith and Arcese 1994). Discriminating the host and parasite is simple once the eggs hatch. Newly hatched cowbirds have distinctive pale gray down and older nestling cowbirds are much larger than sparrows of the same age (figure 2.1i).

Because the cowbird and the sparrow lay similar eggs, identification errors occurred. In a sample of 712 clutches in 7 years with cowbirds, six cowbird chicks hatched from 373 "song sparrow" clutches. Similarly, a sparrow chick hatched from an egg in 1 of 71 hatching and "parasitized" clutches. The low frequency of such errors (7 of 444 clutches, 1.6%), however, means that they had little effect on our analyses.

We checked nest contents carefully near the expected time of fledging (9–11 days after hatching) to estimate the number of young fledged. Subsequent checks of territories confirmed the presence and number of fledged young. Young reach independence from their parents between 24 and 30 days of age (Arcese et al. 2002). Sightings of older fledglings provided information on which individual young survived to independence.

Assigning Gender

Male and female song sparrows are similar in plumage and cannot be sexed reliably as nestlings or juveniles without using genetic markers, which we did not begin to do until the year 2004. Once they become territorially active, males and females behave differently, use different vocalizations, and can be assigned a gender with confidence.

Blood Sampling

Starting in 1987, we collected 30–50 microliters of blood from the brachial vein of each individual for genetic analysis. Adult birds were caught in mist nests set up on trails cut through the band of scrub (figure 2.1b), and young birds were bled in the nest. Some related procedures are described in accounts of our genetic studies (see chapters 6–8).

2.4. Predators and Competitors on Mandarte Island

In addition to the song sparrow, other birds and mammals inhabit the island (Drent et al. 1964). These included 3,000–7,000 breeding seabirds, mostly glaucous-winged gulls, pelagic and double-crested cormorants, and pigeon guillemots. Breeding terrestrial birds include 7–25 breeding pairs of northwestern crows, 2–10 pairs of fox sparrows, and small groups of red-winged blackbirds in some years. Numerous deer mice inhabit the island, particularly the shrub habitat (Merkt 1981).

We know that crows eat sparrow nestlings from the presence of banded sparrow legs at crow feeding perches, and they undoubtedly ate some clutches of eggs. Fox sparrows (40 g) and red-winged blackbirds (45–70 g) are potential competitors of song sparrows, but the species seldom interacted aggressively. In winter, fox sparrows dominated song sparrows at feeders. Up to 1985, spotted towhees visited the island regularly in winter and also dominated song sparrows at feeders (Smith et al. 1980). Since 1985, towhees have not visited the island regularly in winter (P. Arcese personal observations)

The abundance of some other bird species has fluctuated during our study. Crow numbers declined from 13–25 breeding pairs in 1975–1980 (Butler et al. 1984) to 7–10 pairs from 2000–2003. Seabird numbers began to decline about 1980, in association with increasing harassment and predation by bald eagles (Verbeek 1982). Current numbers of all breeding seabirds are less than half of numbers at the start of our study. Canada geese, which were absent up to 1982, have colonized the island and surrounding small islands and now breed commonly (40–45 breeding pairs).

Other species of vertebrates visit the island regularly, and two of them affect the song sparrow. Several thousand starlings roost on the island overnight for much of the summer and fall and defecate millions of blackberry and arbutus seeds annually. These seeds provide a rich and concentrated winter food source for the sparrows (Arcese 1989a) and are contributing to vegetation change on the island. Second, small numbers of brown-headed cowbirds visit the island in some years. Cowbirds influenced the population dynamics of song sparrows quite strongly (see chapter 5). Predatory birds, particularly Cooper's hawks, visited the island regularly and killed some song sparrows (Arcese 1989a; see chapter 4).

2.5. Studies of the Metapopulation

Mandarte Island is isolated by open water from the many nearby islands (figure 2.2). The smaller Halibut Island, 1.3 km distant, is the closest island; it, too, has breeding song sparrows. Thus, Mandarte song sparrows are potentially part of a metapopulation of island subpopulations connected by immigration and emigration. Starting in 1988, but not with the consistent annual effort that we have used on Mandarte, we also studied song sparrows on some nearby islands. In these studies, we used similar methods to explore the movements of adult and juvenile birds, the synchrony and causes of local shifts in numbers and reproductive success, and their patterns of survival (Smith et al. 1996; Wilson and Arcese in press). We do not present these results in detail here, but we do touch on them occasionally.

3 Life History: Patterns of Reproduction and Survival

James N. M. Smith, Amy B. Marr,
and Wesley M. Hochachka

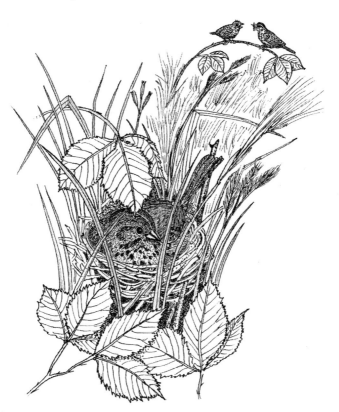

*Female song sparrow wm.o (a white band above a num-
bered metal band on her left leg, and an orange band on her
right leg) was 2 years old and in the prime of her life in*

1981. She laid four clutches of 3, 4, 4, and 2 eggs that year, beginning on March 18 and starting her last clutch on June 14. One offspring survived from each of her first three broods to reach independence from its parents. A sparrow and cowbird fledged from the final nest, and the sparrow survived to at least 17 days of age. Although she was unusual in rearing four broods in a single year, wm.o was in other ways a typical song sparrow, most successful at nesting in her middle ages, producing multiple broods of offspring, and experiencing impacts of cowbird parasitism. These life history patterns underlie many topics covered in the remainder of the book.

Any group of organisms has a characteristic age-dependent pattern of investment in growth, reproduction, and survival—their life history. Our studies over the last 28 years have shown that Mandarte song sparrows are, in many ways, typical songbirds. However, individual sparrows vary greatly around the average life history. In this chapter, we highlight some of these variations, which affect the ecological and genetic dynamics of the population. In chapter 11, we return to the central role of life histories in population biology when we briefly examine how far patterns seen on Mandarte can be generalized to other populations and species.

3.1. Song Sparrows as Typical Passerines

Nesting Biology

Mandarte song sparrows resemble many other passerine birds. During the breeding season, they lay their eggs in open-cup nests. Based on nest type (cavity/crevice vs. open-cup/saucer as classified by Ehrlich et al. 1988), 213 of 258 of North America's passerine species (83%) are like the song sparrow and 17% of species are like the cavity-nesting tree swallow. A further 27 species have domed or sack-like nests; the life history of most of these species has been less studied. Open-cup nesters also dominate in Europe, the Middle East, and North Africa (Cramp 1988, 1992; Cramp and Perrins 1993, 1994a, 1994b).

A key life history trait is the number of sets of offspring produced per female per year. On Mandarte, song sparrows usually start nesting in March and continue through early July, although both the start and

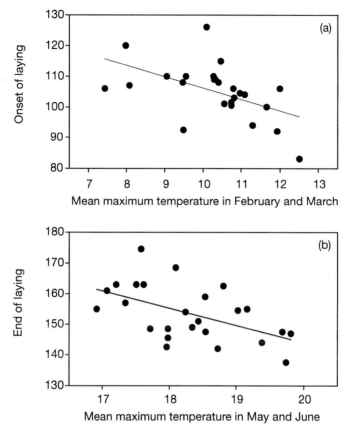

Figure 3.1. (a) Median date of first song sparrow clutches on Mandarte in relation to the mean minimum daily temperatures (°C) in February and March. (b) Median date of last clutches in relation to mean maximum temperature in May and June. Each point represents one year; three years, 1980 (no data), 1985, and 1988 (feeding experiments), are omitted. Calendar days on the y-axes translate to dates as follows: 91 = April 1, 121 = May 1, 152 = June 1, 182 = July 1. Linear regressions: (a) $Y = 143.42 - 3.72X$, $r^2 = 0.271$, $p = 0.008$; (b) $Y = 256.94 + 5.65X$, $r^2 = 0.277$, $p = 0.007$.

end of the breeding season vary considerably among years (figure 3.1). Up to six clutches are completed per year, and 67% of Mandarte females raise two or more broods per year (figure 3.2). Occasionally, pairs even rear four broods in a year. Elsewhere in its range, the song sparrow also has multiple broods and is an early breeder compared to other songbirds (Arcese et al. 2002).

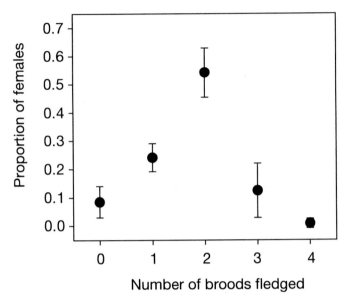

Figure 3.2. Mean proportions of song sparrow females (±95% confidence intervals) that reared 0, 1, 2, 3, and 4 broods to fledging in 12 years when <5% of nests were parasitized by cowbirds.

A typical nesting cycle in the song sparrows on Mandarte takes about 46 days, with 3–5 days of nestbuilding, then one egg laid per day, an incubation period of 12–13 days, a nestling phase of 9–11 days, and juvenile independence 24–30 days after hatching. Reaching 24 days of age was our criterion for assessing survival to independence from parental care in this book. A single nesting cycle is almost as long as the mean interval from laying first to last clutches on Mandarte (48.1 days ± 95% CI = 4.1, n = 27 years). In addition, early nests may fail completely, using up valuable breeding time. Pairs therefore face a time constraint if they are to raise more than one brood of young in a year.

Song sparrow pairs increase their chance of raising two or more broods by overlapping successive breeding attempts. When a brood first fledges, the parents initially divide the brood, with male and female feeding different individual young. Females start to lay a new clutch from 3–20 days after the earlier brood fledges. When the female begins to incubate this new clutch, the male usually takes over most of the feeding of the previous brood (J. N. M. Smith 1978; Smith and Merkt 1980). The bigger the earlier brood is, the more parental care it

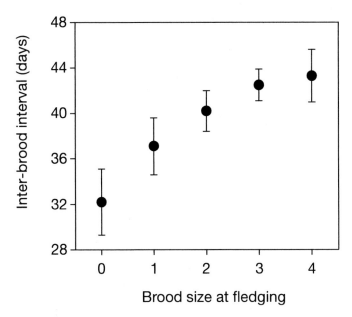

Figure 3.3. Intervals between song sparrow broods in relation to brood size at fledging. The interbrood interval is the number of days between laying of the first egg in the first clutch and the first egg in the subsequent clutch. Error bars are 95% confidence intervals. Data from 12 years when <5% of nests were parasitized by cowbirds.

requires and the longer it takes the female to start a further brood (Smith and Roff 1980; figure 3.3).

Brood division and overlap also occur in other multibrooded north temperate songbirds (e.g., Nolan 1978; Wheelwright et al. 2003) and in some tropical passerines with small clutches and long breeding seasons (e.g., Willis 1972). In the European robin, brood division probably increases the survival of fledglings in a current brood, while also decreasing interbrood intervals (Harper 1985).

Clutch sizes of song sparrows on Mandarte are typical of temperate-nesting passerines with open-cup nests. Four-egg clutches were commonest (48.5%; $n = 714$), followed by clutches of three (44.3%), two (6.6%), five (0.4%), and one (0.3%). Median and mean clutch sizes were four, and $3.42 \pm 95\%$ CI $= 0.04$ ($n = 714$) eggs, respectively. Elsewhere in the song sparrow's large range, clutches are fairly similar in size to those on Mandarte, but exhibit the usual songbird trend of larger clutches in northern and mountain populations (Arcese et al. 2002).

A life history where multiple small clutches are laid per year is common in temperate regions (Gill 1994) and subtropical and tropical regions (Skutch 1954, 1960; Yom-Tov 1987; Yom-Tov et al. 1994). Indeed, it is probably a dominant life history trait among passerines. Another characteristic, exposure to interspecific brood parasites (Davies 2000), is also not unusual. The parasitic brown-headed cowbird removes song sparrow eggs and young, thus lowering clutch size and affecting other aspects of reproductive performance (Arcese et al. 1996; see chapter 5). There were also egg identification errors in years with frequent parasitism (see chapter 2). We therefore described the song sparrow's life history using data from 12 years when less than 5% of clutches were parasitized by cowbirds. This focus allowed us to avoid confounding intrinsic life-history variation in the sparrows with the extrinsically induced effects from cowbirds, which we discuss in chapter 5.

Survival and Age at First Breeding

Song sparrows seem to have typical survival rates for small songbirds, although reliable estimates of survival in the first year of life are extremely scarce. First-year survival on Mandarte averaged 32% ($n = 28$ years) but varied greatly from year to year: Survival was less than 10% during population crashes and reached 82% in 1989–1990. After the first year, annual survival averaged about 60%, except during population crashes, when it, too, dipped to low levels (4–17%). Survival rates of males and females were generally similar overall (see figure 9.1), but females tended to survive less well than males in years with low overall survival (see chapter 4). These rates from Mandarte are higher than most published adult survival rates for small European songbirds (Martin and Clobert 1996). However, some small Australian and South American species have even higher survival rates (Ricklefs 2000; Ghalambor and Martin 2001). Since we could be reasonably sure that any bird that remained alive on Mandarte would be observed, we did not consider resighting uncertainty (Lebreton et al. 1992) when estimating survival rates. No adult sparrows are known to have dispersed away from Mandarte after breeding there.

Virtually all female song sparrows attempt to breed in their second summer (age 1), although some males do not breed until age 2 or later. This tendency of males to begin breeding at older ages also affects average age-specific reproductive success (this chapter), the adult sex ratio (see chapter 4), and the social system (see chapter 6). Delayed onset of breeding by some males is also found in other well-studied passerines (e.g., Beletsky and Orians 1996).

3.2. Patterns of Variability in Life Histories

Intrinsic Sources of Variability

Variability among song sparrows' life histories arises in several ways. These include purely chance events, predictable effects of the rearing environment (see chapter 9), individuals' genetic constitution, and gene–environment interactions (see chapters 7, 8).

Other differences occur within each bird as it ages. Cross-sectional studies of songbirds often reveal that cohorts reproduce more successfully as they age but less successfully near the end of their lives (e.g., Robertson and Rendell 2001; Reid et al. 2003b). Longitudinal studies of individuals also show that reproductive performance falls in old age (e.g., Reid et al. 2003b).

In Mandarte sparrows, annual reproductive success depends strongly on parental age. Numbers of independent offspring and breeding recruits rise sharply with the mother's age up to age 3 and then decline (figure 3.4e,f). There was an 89% increase in numbers recruited from age 1 to 3 (figure 3.4f), but offspring of females 5 years and older were only about as likely to recruit as offspring of age 1 females. Male reproductive performance also increased with advancing age (figure 3.4, open points), again with evidence for declines beyond age 3. Age-dependent performance, however, differed between males and females, particularly among yearlings. Yearling males bred later on average than yearling females (figure 3.4a). As a consequence, they had fewer nests (figure 3.4d), produced fewer fledglings (figure 3.4e), and produced fewer recruits (figure 3.4f) than did first-year females. These differences arose in part because some yearling males only began to breed late after a period as a nonterritorial floater or as an unmated territory owner (see chapter 6). In contrast, virtually all yearling females acquired mates before the beginning of their first breeding season. Once yearling males did gain a mate, however, they performed as well as yearling females when rearing eggs to the fledgling stage (figure 3.4c).

The increasing reproductive performance of song sparrows up to age 3 followed by a decline resembles patterns seen in other songbirds (reviewed in Sæther 1990). Improved success with increasing age is generally attributed to age-related abilities to acquire food and avoid

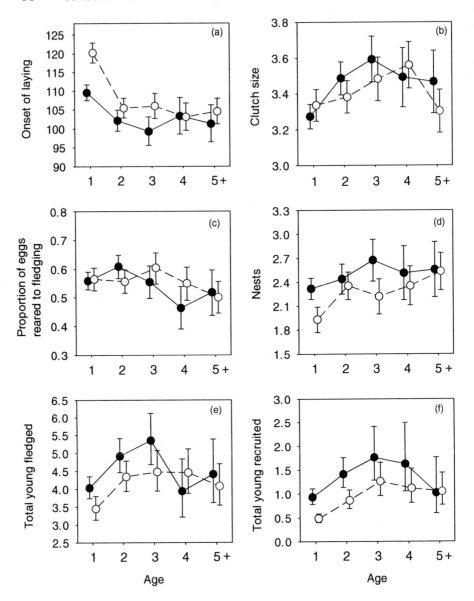

nest predation (e.g., Désrochers and Magrath 1993; Robertson and Rendell 2001). A contributing mechanism for Mandarte sparrows is that early-hatched young, which are more likely to have middle-aged parents, have a higher probability of recruiting than do late-hatched young (Hochachka 1990; see chapter 9). Declines in reproductive performance late in life may also reflect senescence.

Figure 3.4. Age-dependent changes (±95% confidence intervals) in six measures of reproductive performance of song sparrows. Solid points are for females; open points are for males. Data are from 12 years when <5% of nests were parasitized by brown-headed cowbirds. We used generalized linear models (SAS, PROC GENMOD; log link function for nonnormal errors in a, e, and f; logit link function for nonnormal errors in c) in analyses with two categorical variables: year and a combined sex/age variable (2 sexes × 5 ages = 10 classes). Multiple data points from individual birds were included; initial exploration indicated little or no correlation among such repeated data, and thus no corrections for the repeated measures were made. Least-squares means and their associated confidence intervals are plotted. Pre-planned contrasts were used to test for statistical differences in traits between males and females at each age. Different error distributions were assumed as follows (respectively, for a–f): negative binomial, normal, binomial, normal, negative binomial, and negative binomial. Statistical differences ($p \leq 0.05$) between sexes were found at ages 1 and 3 for date of first egg, none for clutch size, none for proportion fledged, ages 1 and 3 for number of nests, ages 1 and 3 for number fledge, and ages 1 and 2 for number of recruits.

◄—————————————————————————————————

Survival in Mandarte song sparrows also depended on age. As have others (e.g., Møller and de Lope 1999), we found declines in survival of older birds, with females generally exhibiting higher mortality (see figure 9.1). As a result, males reached very old age more often than did females (figure 3.5). The oldest individual in the study, a male born in 1992, lived to age 10. The oldest female lived to age 9. However, only four females survived to age 7, compared to 18 males.

Population age structure is a key feature of the demography of long-lived species and remains important in species with shorter lives (Eberhardt 1988). Mandarte's sparrows are no exception, because the age composition of the population varies among years (figure 3.6). Strong reductions in median age occurred after the population crashes in 1979–1980 and 1988–1989. These strong shifts in age structure may have had large demographic consequences, because key aspects of breeding performance varied strongly with age (figure 3.4), but this topic has yet to be studied.

Extrinsic Causes of Variable Life Histories

In addition to the intrinsic variation caused by age, abiotic and biotic environmental disturbances affected individual life histories through

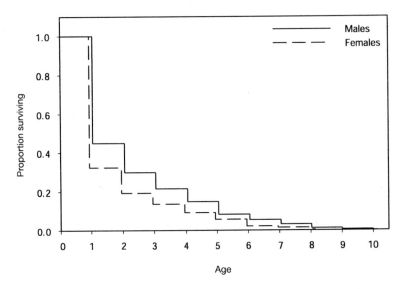

Figure 3.5. Survival curves for male and female song sparrows of varying age on Mandarte Island. Data are for all 28 years.

catastrophic population crashes (see chapter 4), as well as having sub-tler influences on survival and reproduction.

A subtler effect of weather is its influence on the timing of breed-ing of Mandarte's sparrows (Wilson and Arcese 2003). The first nests were begun as early as late February in the warmest El Niño springs, and nesting stopped earlier when May and June were warmer (figure 3.1). We suspect that warm and dry weather in May and June reduced invertebrate numbers in foraging meadows that sparrows used exten-sively in spring. In drier years, there was little use of meadows for for-aging after mid-June, and prey samples from these meadows yielded few invertebrates (J. N. M. Smith, unpublished data).

Much additional variation in the onset of breeding remains after ac-counting for the effects of spring temperature (figure 3.2a). Predators (see chapter 2) and parasitic brown-headed cowbirds (see chapter 5) also affected the survival and nesting success of individual sparrows. Predation on adults was most evident in 1999, when many females were killed in early spring, apparently because they were in poor condition. Diseases and macroparasites have not been systematically studied on Mandarte, but we think that these factors also affected survival rates of birds in 1999, when clutch sizes were unusually small and the on-set of breeding was delayed.

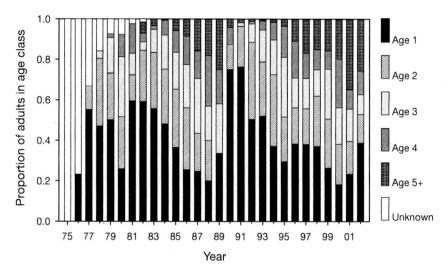

Figure 3.6. Annual variation in age structure of adult song sparrows on Mandarte Island.

These extrinsic influences led to highly variable densities of song sparrows on Mandarte, and sparrow density itself interacted with other factors to determine survival and reproductive rates (see chapters 4, 6, 9).

3.3. Summary

This chapter describes the typical life history of a song sparrow on Mandarte island, a life history that is also typical of many other passerines throughout the world. Variation around the typical patterns has been caused by both intrinsic differences among birds, such as their age, as well as being imposed from outside. Detailed examination of several sources of variation and their consequences, related to the small population size and relative isolation of Mandarte's sparrows (see chapter 2), will form the core of subsequent chapters.

4 Fluctuations in Numbers: Population Regulation and Catastrophic Mortality

James N. M. Smith, Amy B. Marr,
Peter Arcese, and Lukas F. Keller

Each year, we visit Mandarte Island in March to reopen our field camp and census territorial birds just before breeding begins. For 5 years we noted increasing numbers, but when we made our visit in March 1980 we were amazed to find that 88% of the adult song sparrows alive in late summer of 1979 had disappeared, as had a similar fraction of the young born in 1979. We later found catastrophic losses on two other occasions.

All populations have the potential for unlimited increase in the absence of restriction on reproduction or survival (Malthus 1798). One framework in which to discuss these restrictions involves identifying the presence and effect of *density-independent* and *density-dependent* factors, which can each limit population growth (Sinclair 1989). Density-independent limiting factors reduce reproduction or survival with equal force at all population sizes and destabilize populations overall. These effects are often related to environmental variation and lead occasionally to *population crashes* due to episodes of catastrophic mortality. In 1989, Hurricane Hugo killed roughly 67% of the red-cockaded woodpeckers inhabiting Francis Marion National Forest, South Carolina (Hooper et al. 1990). In contrast, density-dependent limiting factors increase in effect as populations grow and thus *regulate* populations at an equilibrium *carrying capacity* in the absence of environmental variation (Sinclair 1989).

In nature, the influence of density-dependent and -independent factors varies by environment and species life history, and they interact to affect population size and stability overall. Populations of short-lived songbirds, for example, often fluctuate irregularly as a consequence of temporal variation in the environment (e.g., Greenwood and Baillie 1991; Sæther et al. 2000b). In contrast, populations of raptorial birds tend to be more stable from year to year because density-dependent regulatory factors dominate their dynamics (Newton 1991). Among Soay sheep on Hirta Island, Scotland, winter weather, population density, and age structure all influence numbers over time (Coulson et al. 2001). It is also the case that some limiting factors have greater effect in combination than alone. For example, combining experimental treatments of food addition and mammalian predator exclusion in a single study plot increased the density of snowshoe hares in the Yukon by 14.4-fold relative to controls, whereas these treatments applied separately caused 3.8- and 2.4-fold increases, respectively (Hodges et al. 2000).

Limiting factors can be further classified as *extrinsic*, acting via an external influence such as weather or competition for food, or *intrinsic*, acting via the physiological or genetic attributes of individuals. Extrinsic limiting factors are observed commonly in bird populations and include inclement weather, shortages of food and nest sites, predators, parasites, diseases, and intraspecific competition for space (Newton 1998). Intrinsic limits on populations may act via the physiological attributes of individuals (Chitty 1967), their degree of inbreeding (Arcese 2003), developmental history (Benton et al. 2001), or age (Gaillard et al. 2000). In this chapter, we first describe the population trajectory of song sparrows on Mandarte Island and then use 28 years of demographic data to explore the effects of limiting and regulating factors. In chapters 6–10, we examine more closely the influence of particular intrinsic and extrinsic effects on population growth.

4.1. Variation in Numbers of Song Sparrows on Mandarte Island

Marked temporal variation in the number of song sparrows present on Mandarte Island from 1975 through 2002 allowed us to explore how survival, reproduction, emigration, and immigration influenced abundance. We use the number of breeders as an index of density because all breeding habitat was defended even at low population size. Food addition experiments were conducted in 1979, 1985, and 1988, but sufficient controls were maintained in each year to describe demography in the absence of added food.

Population Crashes

The dynamics of the Mandarte population were punctuated by precipitous drops in numbers in the winters of 1979–1980 and 1988–1989 and a more gradual decline during 1996–2001 (figure 4.1). Despite their roughly cyclic appearance, several observations suggest that these fluctuations resulted from catastrophic episodes of mortality of varying cause. The precise timing of the 1979 decline is unknown, but it was complete by mid-March 1980 when the first annual census was conducted. Although more extreme weather conditions occurred in the absence of declines (Arcese et al. 1992), December 1979 was the third wettest month in our study, and January 1980 the fifth coolest in terms of mean monthly maximum temperature.

The decline of 1988–1989 was more clearly linked to harsh winter weather. Chris Rogers studied the details of body condition during 1988–1989 and showed that adult and yearling survival was high to mid-January 1989 (Rogers et al. 1991). From January 31 to February 4, 1989, however, mean daily temperature dipped to −11°C, at the lowest point accompanied by northeast winds averaging over 40 km/hr. On February 9, Rogers and W. Hochachka returned to locate a few survivors and one dead adult female with no visible fat. Overall, only 8 of 120 breeders (three female, five male) from 1988 survived the crash,

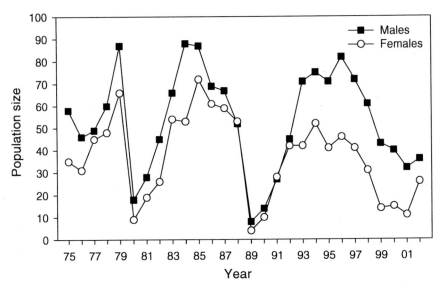

Figure 4.1. Numbers of adult sparrows alive on Mandarte Island in April each year from 1975 through 2002.

and only 3 of over 40 juvenile birds alive in mid-January survived to March.

In contrast to the case on Mandarte, the survival of marked song sparrows on nearby islets and in old-field habitats near Vancouver was about normal in 1988–1989. A potential explanation for this spatial variation in survival is that severe winds during the coldest part of 1988–1989 precluded access to the intertidal areas frequented by birds for feeding (see chapter 2). On other monitored islands, sheltered feeding sites probably remained available to sparrows on their southwest sides, whereas cliffs characterize Mandarte's southwest side (see chapter 2).

A third period of pronounced mortality occurred in late winter 1998–1999, when at least 12 sparrows were killed by accipiter hawks, based on bands recovered at roosts (P. Arcese, unpublished data). Of seven victims that could be aged, five were yearlings and two were 6-year-olds, implying that age or social status may have played a role in mortality. However, on April 7 and 12, two sparrows were found dead, intact and emaciated in their territories. On April 15, a Cooper's hawk thought to have been responsible for several sparrow mortalities was also found dead and emaciated. Because accipiter hawks visited Mandarte regularly and often caused mortality (Arcese 1989a), whereas dead, intact sparrows were rarely observed in any year, we suspect that disease rendered sparrows unusually vulnerable to predation in 1999. Interestingly, the Mandarte sparrows bred 2–3 weeks later than did birds on nearby islands in 1999 (Wilson and Arcese, in press), consistent with the notion of poor general condition.

Tompa (1964, 1971) also documented high mortality in March 1962, coincident with a rare snowstorm. In this case, mortality fell heavily on territorial adults and yearlings. In contrast, 36 nonterritorial juvenile males survived without loss by flocking and feeding on the south side of the island where snow was scarce.

In summary, four episodes of high mortality occurred in mid to late winter; two of them coincided with harsh weather. A third year of high mortality occurred during a cool La Niña in 1999, when birds may have become vulnerable to predation in association with disease. The cause of mortality in 1979–1980 is unknown. These mortality events led to marked fluctuations in population size (figure 4.1) that demonstrate the effects of extrinsic environmental limits on population size. Of 15 populations considered by Lande et al. (2003), the Mandarte song sparrows had the largest coefficient of environmental variance (0.44), exceeding the value for red grouse (0.41), which exhibit cyclic dynamics. This value was also 6.3 times larger than the mean (0.07) of 13 noncyclic species. We now explore what the recovery from these declines might teach us about the influence of density on population growth.

Recovery from Declines

Numbers recovered rapidly after the crashes in 1962, 1979–1980, and 1988–1989. In 1962, nonterritorial males settled at high density on vacated territories, and the number of breeding females rose from 44 pairs in April 1962 to 69 in 1963 (Tompa 1971). After the two most severe crashes, the number of breeding females roughly doubled in each of 3 successive years (mean rate of increase = 1.82 from 1980 to 1983 and 2.19 from 1989 to 1992), propelled by high juvenile survival (Arcese et al. 1992; see figure 3.6). In contrast, numbers remained low after the 1999 decline (figure 4.1), increased in 2002, but declined again thereafter (see chapter 11). The absence of a strong rebound in numbers after 1999 suggests that a change in the presence or effect of factors limiting population size may have occurred on Mandarte.

Rapid recovery from catastrophic decline also occurred in Capricorn silvereyes on Heron Island, Australia. In two of three declines caused by hurricanes, numbers recovered the following year, the third requiring 2 years (McCallum et al. 2000). Soay sheep also rebounded rapidly and repeatedly after winter crashes (Coulson et al. 2001). In contrast, numbers of Darwin's medium ground finch on Daphne Major Island, Galápagos, changed more slowly, declining sharply after successive drought years and only recovering rapidly after wet El Niño years (Grant and Grant 1989; Grant et al. 2000). Overall, the Mandarte Island song sparrows offer an extreme example of island populations of vertebrates subject to marked instability. The rapid recovery of island populations following catastrophic decline suggests that density-dependent limits on reproduction or survival are often relaxed following population decline and the return of favorable environmental conditions.

4.2. Density Dependence

Arcese et al. (1992) showed that adult reproductive rate and juvenile survival rate each increased on Mandarte Island at low population size. More recently, however, the population has varied less predictably, particularly from 1996 to 2001 (figure 4.1). We therefore re-examined the strength of density-dependent reproduction and survival with 14 years of additional data, and also tested for an effect of density on the number of immigrants to the island.

Reproduction

We found that the per capita production of independent young declined at high population density, but also that the strength of this relationship weakened in the latter third of our study (figure 4.2). Because reproductive rate also varied in relation to the annual timing of breeding, clutch size, and nestling and fledgling survival rate (see chapter 3), we asked if these components of reproductive success varied with population density.

We found that the timing of laying first clutches did not vary strongly with density (figure 4.3a), but last clutches were laid earlier at high density (figure 4.3b). Three of the four components of annual reproductive rate also varied with population density. Clutches were smaller (figure 4.3c), fewer eggs survived to hatch (figure 4.3d), and the proportion of fledglings that survived to independence were all lower at high density (figure 4.3f). The proportion of hatched young that fledged was independent of density (figure 4.3e).

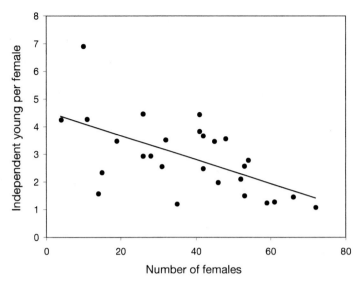

Figure 4.2. Mean number of independent young raised per female in relation to log(female) numbers in spring 1975–2002: $R^2 = 0.35$, $p = 0.001$. In generalized additive models, the spline functions describing the relationship between density and ARS interacted significantly with year of study; that is, density dependence weakened over time.

Several mechanisms might account for the patterns above. Density dependence in clutch size is a common but not universal observation in birds (e.g., table 5.1 in Newton 1998; Both 2000; Krebs 2002) and is often attributed to competition for food prior to or during the laying period (Perrins and McCleery 1994; but see Both 2000). In support of this idea, pairs provided with supplemental food at high density in the spring and summer of 1985 on Mandarte did lay larger clutches than did controls (Arcese and Smith 1988). However, egg removal by brown-headed cowbirds also reduced clutch size at high densities (see chapter 5), and supplemental feeding reduced parasitism (Arcese and Smith 1988). Thus, added food may have acted on clutch size via its effects on parasitism (Arcese and Smith 1988, 1999). A second density-dependent effect, egg loss before hatching, was mainly due to predation or desertion of entire clutches at high densities. We show in chapter 5 that cowbirds also caused many of these losses.

Our observation that breeding ended earlier at high density is also interesting, with a strong potential to affect reproductive output via its

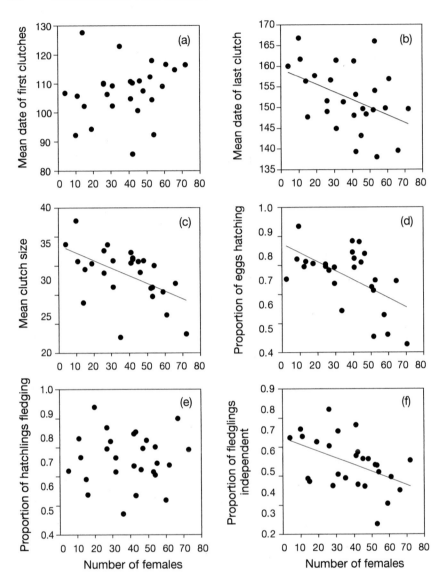

effect on the length of the breeding season. As territory size declined at high density (see chapter 6), it is plausible that birds ran short of the invertebrate foods required for renesting more often.

The reduced survival of fledglings at high density is also a new result that might be overlooked in populations wherein survival from fledging to independence is less easily measured. Three explanations for an increase in fledgling losses at high density might include predation by

Figure 4.3. Correlation between population density and (a) mean laying date of first clutches, (b) mean laying date of last clutches, (c) mean clutch size, (d) mean proportion of eggs raised to hatching, (e) proportion of hatching young raised to fledging, and (f) proportion of fledglings raised to 24 days of age. Regressions for effects of density on log-transformed x-variables: mean date of first clutches (a), $R^2 = .04$, $p = 0.31$; mean date of last clutches (b), $R^2 = 0.19$, $p = 0.023$; mean clutch size (c), $R^2 = 0.27$, $p = 0.005$; proportion of sparrow eggs hatching (d), $R^2 = 0.31$, $p = 0.003$; proportion of hatchlings fledged (e), $R^2 = 0.00$, $p = 0.98$; proportions of fledglings independent (f), $R^2 = 0.22$, $p = 0.013$. The proportion of fledglings surviving to independence varied with female density, and a random effect, the pair of parents. Logistic regression: both $p = 0.001$, $n = 24$ years (1975–1999, excluding 1980). Mean date of the first egg in first clutches and mean hours with rain/day in 7 days postfledging were not significant.

◄───

(1) resident and (2) visiting predators and (3) food shortage mediated by intraspecific competition. Resident predators of fledglings on Mandarte include the northwestern crow (we occasionally found the legs of fledglings below crow feeding perches), glaucous-winged gull (one gull was observed to snatch an adult male sparrow from a song perch in flight), and the deer mouse (we occasionally observed nestlings with bite marks). Daily logbooks for April–July from 1975–1993 also recorded visits by potential predators of fledglings; common ravens visited often (100 records), followed by barn owls (54 visits) and accipiter hawks (17 records). We cannot rule out that one or more of these resident or visiting predators fed more often on fledgling sparrows at high density, but we also have no evidence to suggest that their numbers or habits varied in a way that might lead to density-dependence in fledgling survival.

Fledglings may also have experienced greater food shortage at high densities. In the week after they leave the nest, song sparrow fledglings do not fly well, seek dense cover, and behave cryptically. During food shortages, fledglings that beg more conspicuously might expose themselves to predation, or they may simply starve more often. Data from supplemental feeding experiments during breeding in two high-density years, 1985 and 1988, bear on this hypothesis. In each case, experimental pairs were provided with feeders while control pairs were not (Arcese and Smith 1988). Over these two years, 82% of fledglings from fed pairs survived from fledging to independence compared to 63% of control fledglings.[1] However, because we did not conduct similar experiments in low-density years, it remains possible that supplemental feeding enhances survival at low as well as high density.

The proportion of nests failing per year was also related positively to female density.[2] We explore this relationship further in chapter 5 because of its link to the presence of brown-headed cowbirds.

In summary, annual reproductive success declined as density rose, but the strength of this relationship weakened in the latter part of our study. Five components of reproductive success that varied inversely with density included clutch size, the end of breeding, egg losses, the proportion of nests that failed, and the survival of dependent fledglings.

Survival

Survival in the nonbreeding season has often been identified as the key factor explaining annual losses of birds and frequently declines as population density increases (reviewed in Newton 1998, table 5.1). We observed this pattern for juvenile but not adult song sparrows on Mandarte. Annual survival of juveniles varied strongly from year to year and was lower in years with high female density (figure 4.4). As in the case of annual reproductive success, however, density dependence in juvenile survival weakened in the last third of our study.[3]

The weakening of the relationship between density and juvenile survival occurred mainly among juvenile females. If we assume that half of all juveniles reaching independence were females, the proportion of females that recruited fell from $0.39 \pm SE = 0.06$ before 1990 to 0.32 ± 0.05 after 1989, whereas recruitment in males remained about constant (0.41 ± 0.05 before 1990, 0.40 ± 0.04 after 1989). These differences also contributed to a bias in adult sex ratios that became most pronounced in 1999–2001, when there were 2.91 males per female on average (figure 4.1).

In contrast to the case for juveniles, adult survival was unrelated to the numbers of juvenile or adult competitors (figure 4.5). It is interesting, however, that survival in adult females was about twice as variable (coefficient of variation = 30.4%) in noncrash years below median density than was male survival (coefficient of variation = 14.1%). Although the annual survival rates of males and females were strongly positively correlated in all years ($R^2 = 0.79$), this correlation was reduced with the two severe crash years, 1979–1980 and 1988–1989, excluded from the analysis ($R^2 = 0.25$).

In summary, juvenile survival declined at high density but was less closely related to density in the last third of the study. In contrast, variation in adult survival was independent of population size; despite that all birds survived poorly in two crashes that occurred coincident with

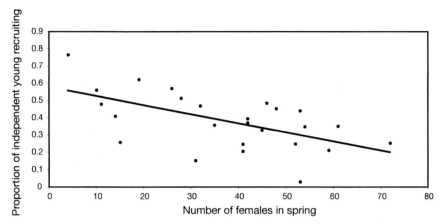

Figure 4.4. Proportion of independent song sparrow young surviving locally to 1 year of age in relation to female density, $R^2 = 0.335$. Regressions of female density on the proportion of adult females surviving, $R^2 = 0.07$, $p = 0.17$; the proportion of adult males surviving, $R^2 = 0.0.07$, $p = 0.19$. The regression of male density on the proportion of adult males surviving gave $R^2 = 0.02$, $p = 0.49$.

high density. Overall, adults survived much better than did juveniles, and juvenile males generally survived better than did females. These results augment those in chapter 3, which show that survival varied with age in adults and declined markedly in very old adults.

Dispersal

The immigration and emigration of individuals are the most difficult aspects of population biology to study (Brooker and Brooker 2002). Emigration may be underestimated if dispersers that leave study areas are not distinguished from those dying within it (Koenig et al. 1996). Immigrants may also be mistaken for neighbors sampled at the edges of a study area. These problems can be reduced, however, where residents are marked in the nest, resighted after fledging, and distinguishable from immigrants.

Dispersal is often density dependent (Lambin et al. 2001), with immigration more frequent at low densities (Massot et al. 1992) and emigration more frequent at high densities (Denno and Peterson 1995). However, more complex patterns can occur. In small mammals, emigration is more common during population increases than during declines (Gaines and McClenaghan 1980), and emigration can be inversely

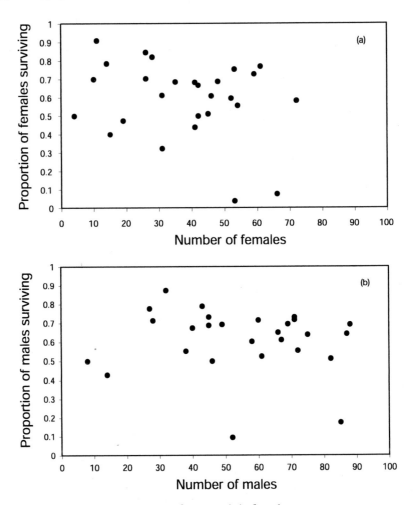

Figure 4.5. Annual proportion of (top) adult female song sparrows surviving in relation to number of females alive the previous spring ($R^2 = 0.071$) and (bottom) adult males surviving in relation to numbers of males alive on the previous spring ($R^2 = 0.018$).

density dependent at low densities, thus increasing the risk of population extinction (e.g., Kuussaari et al. 1996).

We estimated immigration rates to Mandarte by marking all local young as nestlings or dependent fledglings, except in 1980. Unmarked breeders that appeared on the island were therefore defined as yearling immigrants, except in 1981. The only immigrant to Mandarte of known origin was a yearling female, and all but three marked birds known to

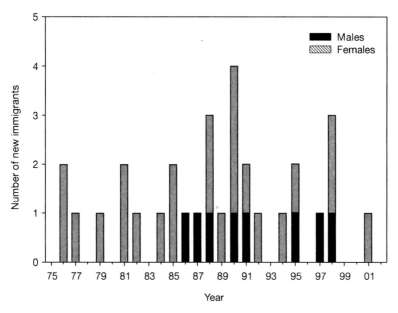

Figure 4.6. Numbers and genders of immigrant song sparrows arriving on Mandarte Island each year. Shaded bars are for females, and solid bars are for males. Data are missing for 1980 because censuses that spring were less complete than in other years.

have dispersed between islands near Mandarte were yearlings (Smith et al. 1996; S. Wilson, A. Marr, and P. Arcese, unpublished data). These immigrants also differed genetically from local birds (see chapters 7, 8).

Over the entire study the number of immigrants averaged 1.2 per year (figure 4.6), and thus had a small influence on annual changes in abundance overall. The maximum number occurred in 1990, when four of 29 breeding adults were immigrants. As in many other birds (Greenwood 1980), female immigrants outnumbered males (74% of 31 immigrants).[4] However, male immigrants survived over twice as long (mean $\pm$ SE life span = 2.25 $\pm$ 0.65 years) as females (0.9 $\pm$ 0.25 years) on average. Surprisingly, the number of immigrants arriving annually was unrelated to population density in the year of recruitment ($R^2 = 0.01$) or the previous year ($R^2 = 0.01$). It is tempting to speculate that the low number of immigrants to Mandarte was due to song sparrows being unable to fly across open water. However, a small number of song sparrows displaced up to 60 km distance from islands in the region of Mandarte crossed at least 7 km of open water to return to their territories. (Arcese 1989a; Smith et al. 1996).

Immigration can also be estimated as a rate per generation. From 1982 through 2002, birds that survived to recruit locally were hatched to mothers 2.0 years old and fathers 2.5 years old, on average, yielding a generation time of about 2.3 years. Thus, the number of immigrants per generation was about 2.8, or 1.2 immigrants per year times the mean generation time. Despite the small overall number of immigrants to Mandarte, however, we show in chapter 10 that they contributed positively to population persistence by augmenting population size at low density.

Although we did not estimate emigration rates, a few banded juveniles are known to have bred on Vancouver Island and other smaller islands nearby (Arcese 1989a; Smith et al. 1996). Tompa (1963) noted the frequent movement of banded young to adjacent Halibut Island in the 1960s, but fewer cases were detected from 1975 through 1986 (surveys of Halibut were rare after 1986). Adult floaters occasionally moved between Mandarte and Halibut islands, but no breeding birds from Mandarte have been detected as emigrants to other islands (Tompa 1963; Smith et al. 1996; S. Wilson and P. Arcese, unpublished data).

Population Growth

Annual change in population size integrates the three main demographic rates, reproduction, survival, and dispersal. We therefore tested if population growth rate, measured as N_{t+1}/N_t, was reduced when female densities were high. Our analyses show that population growth was nonlinear and particularly high at the lowest densities. Growth rate was also higher in years with relatively larger cohorts of independent young.[5] However, four low-density years late in the study, 1999–2003, showed a departure from the overall pattern. In these years, juvenile survival averaged 0.36 and population growth rate 1.30, compared to values of 0.61 and 2.18, respectively, in 5 other years with fewer than 30 females. We now consider this shift in dynamics further.

4.3. Shifting Patterns of Population Regulation

Strong regulation characterized the Mandarte song sparrow population through 1989 despite two crashes in numbers (figure 4.1; Arcese et al. 1992). After 1989, however, the population behaved somewhat differently. First, instead of recovering to the peak densities of 1979 and 1985, numbers stabilized over the 5 years from 1993 through 1997 and then declined to low density in the absence of catastrophic mortality. Second, as noted above, the production of young was less closely correlated to population density late in our study. About 59% of the annual variation in the per capita production of young per female was accounted for statistically by density to 1989, but only 35% of this variation was explained by density over the entire study (figure 4.2). One mechanism affecting these changes is explored further in chapter 5. In brief, brown-headed cowbirds visited the island infrequently after 1989 and were a principal cause of density-dependent reproduction in our study.

In addition, however, the effect of density on juvenile survival also weakened after 1989. Coefficients of determination (R^2) for the relationship between density and juvenile survival fell from 0.74 up to 1989 to 0.35 afterward, with lower survival among juvenile females being the main cause of this change. These reductions in the effect of density on reproduction and juvenile survival rate raise the possibility that a common cause led to temporal variation in the strength of regulating and limiting factors. We revisit these ideas in several subsequent chapters.

In summary, the Mandarte population was strongly regulated by variation in juvenile survival and adult reproductive success over the first 14 years of our study. During the last third of our study, however, these effects weakened. As a consequence, the effect of density on the regulation of population size was less evident over our entire study than during the first 14 years.

4.4. Comparisons of Insular and Mainland Populations

Because more than half of all endangered bird species live on islands (Bell and Merton 2002), and many extinctions of island birds have already occurred (Steadman 1995), the dynamics of island populations are highly relevant to conservation. Many populations also occur in continental habitat "islands." It may therefore be useful to ask if studies of true island populations might be generalized to continental habitat fragments. At least two points, however, suggest caution about such generalizations. First, the dynamics of species vary in space, potentially limiting the application of models based on a single population (e.g., Breininger et al. 1996). Second, limiting factors may differ between true islands and habitat fragments, in part because islands may incur low immigration compared to continental areas (Brown and Kodric-Brown 1977) and support fewer predators (e.g., Arcese and Smith

1999). Colonial seabirds also provide nutrient supplements to some oceanic islands via guano deposition. Thus, island populations may experience less predation and higher quality habitat than patchily distributed continental populations on average (Burke and Nol 1998).

With these differences in mind, we used data from 28 long-term studies of land birds to ask if patterns of variation in numbers differed between continental and island populations (table 4.1). We found that island populations were more variable than mainland ones on average (table 4.1). Eight island populations had a mean ± SE coefficient of variation of 0.42 ± 0.05 versus 0.24 ± 0.03 for 20 mainland populations. The amplitude of variation in population size was also about three times greater for island than mainland populations (6.82 ± 1.22 vs. 2.23 0.87, respectively; table 4.1). In general, populations with higher coefficients of variation and amplitudes had also been studied for longer.

4.5. Discussion

Density Dependence, Demographic, and Environmental Variability

Our study, and work on Capricorn silvereyes (McCallum et al. 2000), illustrates the vulnerability of small, isolated populations to environ-

Table 4.1. Coefficients of variation and amplitudes (maximum N/minimum N) of population size for land birds inhabiting small islands and mainland sites.

Island Population/Species (Source)	Years of Data	Coefficient of Variation	Amplitude
Capricorn silvereye (1)	27	0.16	2.1
Medium ground finch (2)	19	0.85	11.0
Cactus finch (2)	19	0.49	9.0
Large cactus finch (3)	8	0.55	4.7
Rock pipit (4)	9	0.25	2.2
Galapagos mockingbird (5)	10	0.30	2.9
Savannah sparrow (6)	16	0.15	1.8
Song sparrow (this study)	27	0.48	18.0
Island Mean (±SE)		0.40 (0.08)	6.46 (2.05)
Mainland Population/Species			
Florida scrub jay: prime habitat (7)	17	0.25	2.05
Florida scrub jay: sparse habitat (7)	17	0.13	1.48

Table 4.1. *(Continued)*

Island Population/Species (Source)	Years of Data	Coefficient of Variation	Amplitude
Black throated blue warbler (8)	30	0.30	4.00
Great tit: Wytham (9)	15	0.23	2.13
Great tit: Oosterhout (10)	21	0.24	2.35
Great tit: Liesbos (10)	22	0.24	2.29
Great tit: Antwerp (11)	35	0.30	3.41
Great tit: Saxony (18)	20	0.17	—
Blue tit: Wytham (18)	37	0.28	—
Blue tit: Antwerp (11)	35	0.26	3.50
Blue tit: Saxony (18)	20	0.27	—
Black-capped chickadee (12)	10	0.11	1.33
Dunnock (13)	9	0.12	1.35
Pied flycatcher: UK (14)	17	0.22	2.11
Pied flycatcher: Saxony (18)	20	0.22	—
Pied flycatcher: Lingen (18)	20	0.25	—
Pied flycatcher: Finland (18)	20	0.45	—
Red-winged blackbird (15)	16	0.22	2.25
European dipper (16)	20	0.49	4.33
European sparrowhawk (17)	13	0.10	1.43

Note. All studies had eight or more estimates of annual population size and were unaffected by temporal changes in management or methods, such as changes in the supply or design of nest sites. Populations exhibiting long-term cycles were also excluded. We used the total number of breeders annually to estimate degree of fluctuation, or the number alive at the normal time of breeding for species such as Darwin's medium ground finch that skip breeding in unfavorable years (Grant and Grant 1996b). We used the coefficient of variation (CV) in population size as an index of variability and calculated the maximum amplitude of fluctuations (maximum N/minimum N) for 22 populations. Since measures of variation are liable to increase with length of study, the number of study years was included as a covariate in a general linear model (GLM) testing the effects of island vs. mainland status on $\log10(CV)$ (island vs. mainland: $F_{1,19} = 6.85$, $p = 0.015$; year: $F_{1,19} = 3.28$, $p = 0.08$) and $\log10(amplitude)$ (island vs. mainland, $F_{1,19} = 12.59$, $p = 0.002$; year, $F_{1,19} = 8.12$, $p = 0.01$).

Sources of data are as follows: 1, McCallum et al. (2000), Brook and Kikkawa (1998); 2, Grant and Grant (1996b, unpublished data); 3, Grant and Grant (1989); 4, Askenmo and Neergaard (1990); 5, Curry and Grant (1989); 6, N. Wheelwright (unpublished data); 7, Woolfenden and Fitzpatrick (1991); 8, Holmes and Sherry (2001); 9, McCleery and Perrins (1991), data from 1960 to 1973 only—changes to methodology affected later estimates of population size; 10, van Balen and Potting (1990); 11, Lande et al. (2003); 12, Smith (1991); 13, Davies (1992); 14, Harvey et al. (1988); 15, Beletsky and Orians (1996); 16, Sæther et al. (2000a); 17, Newton (1998); 18, tables 1.1 and 1.2 in Lande et al. (2003).

mentally induced population crashes. Such crashes brought song sparrows on Mandarte to very low numbers (9, 4, and 11 females) three times in 28 years, such that extinction due to stochastic variation in demography became likely (see chapter 10). Following two of these crashes, numbers recovered to at least median density over the following 3 years. On the third occasion, numbers did not recover until the fourth year, and later declined again (P. Arcese, unpublished results; see chapter 7).

The first two recoveries from low numbers were caused by strong density dependence in population growth, via improved reproductive success and increased juvenile survival at low density. Because density-dependent juvenile survival operates after density influences annual reproductive rate, it is the dominant mechanism of regulation in the population (Arcese et al. 1992). Capricorn silvereye numbers also recovered twice within a year after episodes of mortality in hurricanes. (McCallum et al. 2000). Strong density dependence in mechanisms affecting population growth is encouraging from a conservation perspective, because populations that spend less time at low densities are also less prone to extinction via the effects of random demographic variation in survival, fecundity, or sex ratio.

Could we use these findings to manage the Mandarte population's long-term persistence? Based on results to 1992, one might have argued that regulatory mechanisms were sufficiently robust that management was not needed to rebuild the population following a crash. Similarly, McCallum et al. (2000) argued that density-independent population viability models may generate unduly pessimistic predictions about extinction probability. However, the failure of the Mandarte population to rebound rapidly from low density after 1999 suggests that supplementing feeding (Smith et al. 1980; Arcese and Smith 1988) or translocations to alleviate inbreeding (see chapter 8) might have hastened its recovery. In chapter 10, we show that even small supplements to population size at low density can reduce extinction probability markedly.

Models of the dynamics of the Mandarte population up to 1998 noted that density dependence in the population was strong, but that its effects were uncertain due to high demographic and environmental variability (Tufto et al. 2000; Sæther et al. 2000a). Indeed, Mandarte song sparrows had the highest coefficient of environmental variability of 15 bird populations described by Lande et al. (2003). The poor growth rates at low densities in the last third of our study add to this uncertainty. Additional factors affecting the persistence of the Mandarte population are considered further in chapter 10.

Temporal Changes in Population Performance

Our results suggest that even extended studies will not always provide a firm basis for predicting the performance of populations in the future. Indeed, a general rule might be that the longer studies continue the more likely it becomes that populations will experience changes in their physical or biotic environment that alter their dynamics. As our study progressed, strong regulatory mechanisms observed in the first half of our study, related to annual reproductive success and juvenile recruitment, became less closely linked to population density in the later part of our study. We now consider if these temporal changes in demography may also have influenced population growth rate.

Two changes in the dynamics of the Mandarte population coincided with reduced population growth during the 1990s. First, recruitment of juvenile females failed to rebound at low density, leading to strongly biased adult sex ratios after 1993 and many unmated males (see chapter 6). In the most extreme year, 1999, only 14 of 57 adults were females. However, the reasons for low female recruitment remain unclear. It is possible that vegetation change (see chapter 2) reduced the survival of juvenile females by reducing cover and increasing their exposure to avian predators, disease, or food shortage or by increasing their rate of emigration from the island. Juvenile females occupy the bottom of an intraspecific dominance hierarchy in winter (see chapter 6) and become susceptible to mortality or emigration when resources become limiting (Arcese 1989a). A second change in the dynamics of the populations involved brown-headed cowbirds, which depress reproductive rate and population growth but visited the island less often after 1992 (see chapter 5). The reduced presence of cowbirds after 1992 may account for the weakening of density-dependent reproductive success, but it did not have the expected effect of improving reproductive success. At present, therefore, we are left with the possibility that an intrinsic effect, such as the average degree of inbreeding, might have reduced population performance in the latter part of our study (see chapter 7).

Other studies of birds have also found marked shifts in population dynamics over time. In the 17th year of a study of Darwin's medium ground finches, numbers increased to densities more than double those encountered during the first 16 years of study (Grant and Grant 1996b; Grant et al. 2000). In this case, changes were driven by multiyear patterns of rainfall and abundant food (Grant et al. 2000). European dipper numbers in Norway also doubled following a series of warmer winters, after an 11-year period of relative stability (Sæther et al. 2000b). Numbers of the endangered Kirtland's warblers remained stable for 18 years and then increased nearly 4-fold over the next 6 years (DeCapita 2000) in part because of habitat changes induced by fire. Forest succession is a familiar directional influence on population dynamics in passerines (Holmes and Sherry 2001), blue grouse (Zwickel 1992), and spruce grouse (Boag and Schroeder 1987). Also, human influences on all populations are becoming more pervasive, even in areas far from civilization (e.g., Stirling et al. 2004). Albon et al. (2000) and Krebs (2002) provide other examples where the effects of density on populations of the same species vary in time and space.

Comparisons of Island and Mainland Populations

Our survey of 28 population studies revealed the relative instability of island versus mainland bird populations (table 4.1). High immigration rates into continental populations may help to dampen their fluctuations. Four mainland studies (Van Balen 1980; Clobert et al. 1988; Davies 1992; Sæther et al. 2000b) revealed that immigrants typically comprise a substantial fraction of new recruits. Seventy-seven percent of recruiting dunnocks were from outside the study area (Davies 1992), 33% of European dippers (Tufto et al. 2000), and about 50% in each of two great tit populations (Van Balen 1980; Clobert et al. 1988). In contrast, most island populations received fewer immigrants, particularly when they were far offshore. On Vlieland Island, Netherlands, 6 km from neighboring Texel, about 20% of recruiting great tits were immigrants (Verhulst and van Eck 1996). Immigrant rock pipits in Sweden made up 66% of recruits on Malön Island, <1 km from the mainland coast, but only 18% on Nidingen Island 6 km offshore (Askenmo and Neergaard 1990). In our study, only 3% of 937 recruiting song sparrows were immigrants. The low proportion of immigrants on Mandarte (1.3 km isolation) is noteworthy, compared to the higher proportion of immigrants on the more distant Vlieland (20%) and Nidingen islands (18%). Presumably, song sparrows are more reluctant to disperse over water than are great tits or water pipits.

Population instability on islands is not restricted to birds. Large mammals, such as Soay sheep (Milner et al. 1999; Coulson et al. 2001), red deer (Clutton-Brock, and Coulson 2002) and caribou (reindeer, Leader-Williams 1988), also exhibit strong fluctuations and frequent population crashes on small islands. Patterns of population change are less well known in other island and mainland taxa but may differ from those for birds and large mammals because of differences in dispersal, life history, and the dominant mechanisms of population regulation. For example, insect populations show higher degrees of local extinction than birds in both continental (Parmesan 1996) and island situations (Hanski et al. 2002).

In summary, bird populations appear to fluctuate more widely in size on islands than mainland areas (table 4.1), a fact that may contribute to their more frequent extinction. In general, immigration is also reduced on island compared to mainland populations. Last, the simpler ecology of small islands may reduce the number of factors limiting population size.

4.6. Conclusions

Mandarte Island song sparrows are regulated mainly by density-dependent juvenile survival, rather than by immigration, which is very limited. Population size was also influenced strongly by the occurrence of population crashes that occurred at roughly 10-year intervals, although not a clearly cyclic pattern of dynamics. Density-dependent juvenile survival and reproductive success were each detected, but both also became weaker during the third decade of our study, when the population appeared unable to recover from decline. A shift in sex ratios, related to the poor survival of juvenile females, accompanied this change in demographic performance. One other small and isolated bird population, of Capricorn silvereyes on Heron Island, Australia, exhibits a similar combination of regulatory effects and catastrophic mortality events. On Mandarte Island, but not Heron Island, catastrophic mortality frequently brings the population to the brink of local extinction. If these patterns apply generally to insular populations, they suggest that augmenting populations at very low density may be necessary to manage the recovery of rare species subject to stochastic variation in the environment.

Notes

1. Mixed model logistic regression (GLIMMIX, SAS) used to test if fledgling survival varied with feeding and year in 1985 and 1988; effect of feeding significant ($F = 9.73$, $p = 0.002$), with no effect of year ($F = 0.13$, $p = 0.72$) or the interaction of treatment and year. Mean survival of fed young was 0.82 (95% confidence limits, 0.76–0.87) and 0.63 for control young (95% confidence limits, 0.50–0.74). Losses within a brood were statistically independent,

as judged by nonsignificance of the random effect "brood identity." The model without this effect explained about 76% of the variance in the data.

2. Regression of log proportion of nests failing per year on density ($R^2 = 0.182$, $p = 0.027$).

3. Regressions of density on proportion of juveniles surviving to breeding age over three periods: $R^2 = 0.736$, $p < 0.001$ (1975–1989); $R^2 = 0.353$, $p = 0.05$ (1990–2001); $R^2 = 0.467$, $p = 0.004$ (1975–2001). Analyses using generalized additive models (GAM) revealed that density dependence in juvenile survival varied significantly with year of study ($p < 0.0001$); that is, density dependence weakened over time.

4. Binomial test, $p < 0.05$.

5. Effect of density on population growth rate as follows: females in year $(t + 1)$/females in year t = adults + adults^2 + juvenile cohort size, where "adults" equaled total alive at the start of breeding. Overall model: $R^2 = 0.69$, $p = 0.001$, $n = 26$ years. Partial R^2 for adults $= 0.32$, $p < 0.001$; that is, higher densities reduced growth. Partial R^2 for adults$^2 = 0.17$, $p = 0.002$; that is, growth rate increased faster at the lowest densities. Partial R^2 for juvenile cohort size $= 0.12$, $p = 0.008$; that is, the larger juvenile cohorts associated with higher growth rate.

5

The Song Sparrow and the Brown-Headed Cowbird

James N. M. Smith, Amy B. Marr,
Peter Arcese, and Lukas F. Keller

When brown-headed cowbirds began to visit Mandarte Island each spring, we saw a sharp increase in rates of nest failure. We noted cases where eggs were removed from nests,

whole clutches disappeared, and nestlings were pecked to death or dragged from the nest. The sketch above depicts an attack by a female cowbird on a song sparrow brood. It is based on an incident videotaped in Victoria, British Columbia, by Anne Duncan.

5.1. The Life History of the Cowbird and Other Brood Parasites

The brown-headed cowbird is one of 100 species of obligate brood parasitic birds worldwide (Davies 2000; Lowther 1993). These birds lay all their eggs in the nests of other "host" species, which then incubate the parasite eggs and rear the resulting young (Rothstein and Robinson 1998a). In one-third of brood parasitic species, including the cowbird, the parasite nestlings are reared together with the host chicks. In the other species, the parasite nestling kills the host nestlings or evicts all host eggs and chicks from the nest (Davies 2000). Brood parasites are often highly mobile and often have distinct feeding, breeding, and roosting areas several kilometers apart (Davies 2000; Curson et al. 2000). In contrast, songbirds with parental care must stay near one place to perform their parental duties and defend their nests and mates.

Brood parasites vary greatly in their dependence on particular hosts. Some species, like many African *Vidua* finches, parasitize only a single host species (Payne 1997). The common cuckoo usually has three or four main hosts (Davies 2000). In contrast, the North American brown-headed cowbird and the South American shiny cowbird each have more than 200 host species (Friedmann and Kiff 1985; Davies 2000). These contrasting host relations influence host population dynamics (see section 5.2).

Brown-headed cowbirds commonly parasitize open-cup nesting songbirds smaller than themselves. Only species that feed their nestlings on insects and other arthropods can successfully rear cowbird young (Rothstein and Robinson 1998b). Female cowbirds often search for host nests from conspicuous perches (e.g., Hauber and Russo 2000; Saunders et al. 2003), using nest building by the host as a cue (McLaren and Sealy 2003), or other active searching methods (Davies 2000).

Once a suitable nest is found, cowbirds visit it at dawn to lay their eggs (Sealy et al. 2000). Most cowbird eggs are laid while the host is laying, and those laid earlier or later often fail to hatch (Nolan 1978;

Davies 2000). Cowbirds' eggs have a spotted appearance common to many ground-nesting songbirds and do not mimic those of any particular host (see figure 2.1h). Some cowbird hosts, however, recognize and eject all cowbird eggs. Most hosts either accept or reject all parasitic eggs, but a few species exhibit intermediate frequencies of egg rejection (Rothstein 1975).

Brood parasites are generally thought to have a high annual fecundity as part of a life-history trade-off against low parental investment (Smith and Rothstein 2000). Annual fecundity varies from 2 to 25 eggs per year in common cuckoos (Davies 2000), to more than 40 eggs per year in brown-headed cowbirds (Scott and Ankney 1983; Smith and Arcese 1994), and as many as 120 eggs in the shiny cowbird (Kattan 1997). However, recent genetic studies have yielded lower estimates (1–13 eggs or nestlings per year) in brown-headed cowbirds (e.g., Strausberger and Ashley 2003; Woolfenden et al. 2003).

Newly hatched cowbird nestlings have two advantages over their brood mates. First, cowbird eggs develop faster than host eggs (Briskie and Sealy 1990). Second, hatching of host eggs is delayed when the host egg is smaller than the cowbird egg (McMaster and Sealy 1998). This 1–2 day head start, and the large size of the cowbird chick, often gives it a competitive advantage. In small host species, many host chicks therefore starve in broods containing parasite nestlings. However, some larger hosts can rear cowbird nestlings and all their own young with few losses (Smith and Arcese 1994; Trine et al. 2000).

In this chapter, we discuss the demographic effects of brood parasitic birds on their hosts and explain why brown-headed cowbirds are of conservation interest in North America. We then describe the demographic effects of cowbirds on song sparrows on Mandarte Island. Finally, we explore the mechanisms underlying these effects and discuss the status of brown-headed and shiny cowbirds as conservation threats.

5.2. Brood Parasites, Host Demography, and Conservation Biology

Populations of brood parasites that specialize on a single host should be limited by negative feedbacks between host and parasite numbers (May and Robinson 1985; Takasu et al. 1993). As expected, most specialist Old World brood parasites are rare. In contrast, host generalists like brown-headed and shiny cowbirds can become abundant and depress the reproductive success of host species and individuals (Trine et al. 1998).

Brown-headed cowbirds are a conservation threat in North America mainly because of their spectacular range expansion (Rothstein 1994; Peterjohn et al. 2000). While expanding their range, cowbirds have encountered new host species, some of which have declined under frequent parasitism (Rothstein and Robinson 1998b). Although cowbirds were recognized more than 70 years ago as a possible cause of songbird population declines (Friedmann 1929; Nice 1937), there remains disagreement over the severity of the cowbird threat (e.g., Ortega 1998; Griffith and Griffith 2000).

Brood parasitism lowers host reproductive success through several mechanisms (Arcese et al. 1996; Smith et al. 2003). Some hosts desert parasitized clutches (Strausberger and Burhans 2001), and parasites often remove host eggs (Scott et al. 1992). Competition with parasite young lowers nestling, and perhaps fledgling, survival. In addition, female brood parasites including brown-headed cowbirds sometimes destroy host clutches and broods (Arcese et al. 1996; Granfors et al. 2001). Chance (1940; pp. 30–35 in Davies 2000), suggested that this behavior in common cuckoos increased the supply of host nests that they could then parasitize. Assuming that female parasites find many host nests too late for them to be used for parasitic laying, destroying such clutches may induce the host to renest nearby, providing a new opportunity for parasitism. As a consequence of these negative effects on reproductive success, brood parasites may depress host population growth rates (e.g., May and Robinson 1985) and reduce population size (see chapter 10).

By the late 1980s, links between high nest failure rates in song sparrows on Mandarte Island and frequent laying by cowbirds became clear (Arcese and Smith 1988). Chance's (1940) hypothesis raised the possibility that host nest destruction is also a reproductive tactic that facilitates parasitism by female cowbirds, particularly when females defend access to hosts (Arcese et al. 1996), as cowbirds seem to do on Mandarte (Smith and Arcese 1994). Although this hypothesis also applies to other brood parasites, here we have called it the *cowbird predation hypothesis*.

While it is obvious that brood parasites can harm host reproduction, it is less clear how much harm they do. A common approach to this question is to compare production of young from parasitized and unparasitized nests in the same population (Trine et al. 1998; Lorenzana and Sealy 1999), sometimes adjusting for the number of broods reared annually (Pease and Grzybowski 1995). Another approach is to follow the reproductive history of parasitized and unparasitized host females over a season (e.g., Sedgwick and Iko 1999; Whitfield et al. 1999) and to also record their and the subsequent survival of their off-spring. A difficulty in these approaches, however, is raised when the parasite destroys host eggs or young. Unless photographic monitoring of the identity of nest predators is used (Granfors et al. 2001; Stake and Cimprich 2003), such attacks may go unrecognized.

Two better ways to estimate the effects of a brood parasite on its hosts are to compare situations where the parasite is naturally present or absent, or to manipulate parasites experimentally (Arcese and Smith 1999). To adopt the first approach, we compared years with and with-out parasitism on Mandarte Island. Cowbirds bred there commonly in only 15 of 27 study years. Experiments to estimate the influence of cowbirds on reproductive success were conducted on the nearby Fraser River estuary (see below).

5.3. Cowbirds and Song Sparrows on Mandarte Island

Cowbirds colonized Vancouver Island in the early 1950s and Mandarte Island between 1963 and 1975 (Tompa 1964; Smith and Arcese 1994). Cowbirds are migratory in the region, and the song sparrow is a frequent cowbird host throughout their joint range (Smith and Myers-Smith 1998). Cowbirds of the Pacific Coast *obscurus* race are bigger than song sparrows of the *morphna* race in coastal British Columbia (adults 30–42 g vs. 23–26 g, respectively), and their nestlings are about 40% heavier at 6 days of age than their song sparrow nest mates (see figure 2.1i).

Methods

We generally saw cowbirds on the island in the mornings, when they flew up and down the island and perched on emergent snags. The pres-ence of a few female cowbirds banded as nestlings on Mandarte, or identifiable by plumage, revealed that one to three cowbird females used the island for breeding annually (Smith and Arcese 1994). Song sparrows recognized cowbirds and mobbed and chased them (Smith et al. 1984; Arcese et al. 1996).

We documented breeding by cowbirds by counting their eggs or chicks in song sparrow nests and estimated their arrival each year from the date that the first cowbird egg was laid. We also described the state of nests, eggs, and nestlings after their failure. Cowbirds and song spar-rows lay similar eggs (figure 2.1h), but their nestlings are easily distin-

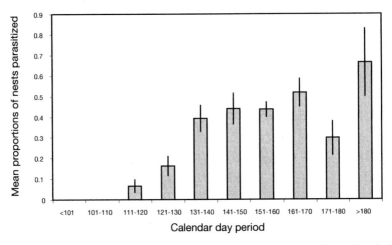

Figure 5.1. Mean proportions of song sparrow nests on Mandarte Island that were parasitized by cowbirds in each 10-day interval: data from 15 years when more than 5% of nests were parasitized. Day 121 = May 1, Day 152 = June 1. Error bars = ±1 SE.

guished (see chapter 2). We also removed all laying cowbirds from the island in 1977 (Smith 1981a).

Annual and Seasonal Patterns of Parasitism

Cowbirds parasitized song sparrow nests in 12 of 13 years up to 1989, but only in six of 13 years thereafter. On average, 14.7% of nests (SE = 0.03) were parasitized annually. We excluded 1977 from these values because all female cowbirds were removed experimentally that year. In 15 of 27 years, at least 13% of sparrow nests were parasitized ("cowbird years"). In 9 of 27 years (including 1977), no nests were parasitized ("years without cowbirds"). In the three remaining years, 1990, 1992, and 1995, only two, one, and three nests were parasitized each year. We pooled these 3 years with years without cowbirds in analyses here.

Cowbirds began breeding later than song sparrows in all years they were present; the first cowbird egg was laid an average of 37.9 (SE = 4.8) days after the first sparrow egg. This time difference meant that the first nests of most female sparrows escaped parasitism. Over all cowbird years, the proportion of parasitized nests rose steeply from zero before April 21 to 39% of nests from May 11 through May 20. The frequency of parasitism then remained stable at about 40–45% of nests until the end of breeding in late June (figure 5.1).

Table 5.1. Mean values ± standard error (SE) for some life history traits of song sparrows on Mandarte Island in years with and without cowbirds, and in relation to parasitism of nests by cowbirds.

| | | Years With Cowbirds | | |
| | | Unparasitized Clutches | | |
Life History Trait	Years Without Cowbirds	Outside Cowbird Laying	During Cowbird Laying	Parasitized Clutches
N (nests)	815	423	506	316
Clutch size	3.35 ± 0.03	3.38 ± 0.03	3.04 ± 0.05	2.42 ± 0.06
No. Hatching	2.61 ± 0.05	2.45 ± 0.06	1.77 ± 0.07	1.40 ± 0.07
No. Fledging	1.94 ± 0.05	1.64 ± 0.07	1.29 ± 0.06	0.97 ± 0.06
No. Surviving to 24 days	1.50 ± 0.04	1.17 ± 0.04	0.88 ± 0.05	0.69 ± 0.05
Proportion of Nests Failing	0.25	0.35	0.43	0.40

Note. Clutches of females with supplemental food in 1989, 1985, and 1988 are omitted. When calculating values outside the cowbird laying period, we used only nests that were initiated more than a week before the first cowbird egg was laid and later than a week after the last egg was laid. Fledglings were judged to be independent of parental care at 24 days of age.

Effects of Cowbirds on Reproductive Success of Sparrows

We estimated the influence of cowbirds on reproduction by sparrows in three ways. First, because cowbirds bred later than sparrows, we compared reproductive performance before and after the onset of cowbird laying. Second, some females in each cowbird year escaped parasitism, allowing us to compare the success per nest and annual reproductive success (ARS) of parasitized and unparasitized females. Third, we compared reproduction in years with and without cowbirds.

We estimated how many eggs were removed by cowbirds by comparing the sizes of parasitized and unparasitized clutches within and across years (table 5.1). Mean sparrow clutch size in years without cowbirds was 3.35 ± 0.03 eggs, approximately one egg larger than in parasitized clutches (2.42 ± 0.06 eggs). Similar differences were seen between clutches laid inside and outside the cowbird laying period within the same years (table 5.1). Parasitized clutches also contained 1.08 ±

0.01 cowbird eggs per clutch for a mean total clutch size of 3.50 ± 0.05 eggs. Unparasitized clutches were also smaller in cowbird years (3.04 ± 0.05 eggs) than in years without cowbirds (3.35 eggs). These values suggest that cowbirds on average removed one egg from each clutch they parasitized, and about one egg from every three clutches that they did not lay in (Arcese and Smith 1999).

Losses of eggs and young in years with and without cowbirds differed mainly prior to hatching. In all cowbird years, 36% of sparrow eggs failed to hatch compared to 22% in all years without cowbirds. Losses of nestlings and fledglings however, were more similar in years with than without cowbirds (nestlings, 30% vs. 26%; fledglings, 30% vs. 23%).

Reproductive success in sparrows differed among years and in relation to female age (see chapters 3, 4). When we corrected for these year and age effects, numbers of fledglings reared per nest were lower for parasitized females (mean ± SE = 1.09 ± 0.13, n = 207 nests) than for unparasitized females (1.76 ± 0.17, n = 248 nests). Parasitized females also reared fewer independent young per nest (0.74 ± 0.12) than did unparasitized females (1.27 ± 0.14).[1] Thus, parasitism caused losses of about 38% of fledglings and 42% of independent young per nest.

Similarly, being parasitized at least once reduced the ARS of parasitized females in cowbird years. After correcting for the effects of year and female age, parasitized females reared 28% fewer fledglings per year (2.77 ± 0.35, n = 248) than did unparasitized females (3.87 ± 0.42, n = 207) and 29% fewer independent young per year (2.22 ± 0.16, n = 248 vs. 3.13 ± 0.27, n = 207).[1]

Parasitism reduced the reproductive success of individual sparrows for two main reasons. First, cowbird nestlings were substituted for sparrows in successful parasitized nests (i.e., those producing at least one fledgling of either species). In cowbird years, successful parasitized nests produced an average of 1.56 fledgling sparrows and 0.66 cowbird fledglings, for a total of 2.22 fledged young per nest (n = 200 nests). Successful unparasitized nests in the same part of the breeding season produced the almost identical average of 2.24 fledglings per nest (n = 269), but all these fledglings were sparrows. Second, more nests failed completely in years with cowbirds.

Cowbirds and Density-dependent Reproductive Success

Cowbirds visited Mandarte Island more frequently and laid more eggs in years with higher densities of female sparrows (figure 5.2). Such *func-*

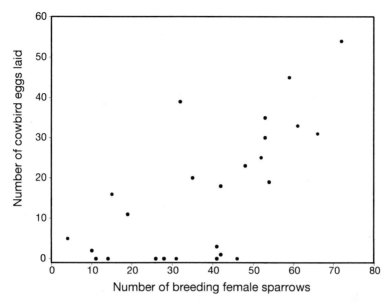

Figure 5.2. Numbers of cowbird eggs laid per year in relation to numbers of breeding female song sparrows.

tional responses may be linear (type 1), saturating (type 2), or sigmoidal (type 3; see figure 13.15 in Krebs 2001). Type 3 responses tend to stabilize interactions of predators and prey, and of brood parasites and hosts. A type 2 curve clearly did not fit the data in figure 5.2, while a type 3 curve (i.e., a linear plus a quadratic term, $R^2 = 0.59$) fit was better than a linear one ($R^2 = 0.49$).

These results imply that cowbirds are involved in density-dependent song sparrow reproduction (see figures 4.2, 4.3). Indeed, density-dependent nest failure was strong in years with cowbirds (figure 5.3a) but absent in years without them (figure 5.3b).[2]

Could cowbirds explain all of the density-dependent ARS documented in chapter 4? To explore this question, we used partial regression analysis to explore how well female density and the proportion of nests parasitized by cowbirds predicted ARS. Both variables contributed to variation in ARS among years, but the effect of density (partial $R^2 = 0.24$) was statistically stronger than the effect of cowbirds (partial $R^2 = 0.10$).[3] This overall model, however, concealed different effects of density and parasitism on components of ARS.

Six main components influence ARS in song sparrows: the mean start of laying first and last clutches, mean clutch size, the proportion of eggs that hatch, the proportion of hatching young that leave the nest

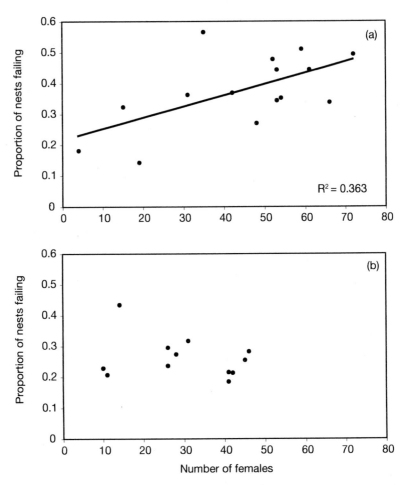

Figure 5.3. Density dependence of rates of nest failure in song sparrows in years with cowbirds (a) and years without cowbirds (b). $R^2 = 0.081$ for years without cowbirds.

successfully, and the proportion of fledglings that survive to 24 days of age (see chapter 4). We again used partial regression to explore how each component was predicted by both density and parasitism. The timing of first clutches was unrelated to density or parasitism. Last clutches were laid earlier when densities were high, but independently of parasitism. Clutches were smaller when more nests were parasitized, but not significantly smaller at higher densities. Similarly, the proportion of eggs hatching was lower with more parasitism but was unaffected by density. Neither parasitism nor density affected the propor-

Table 5.2. Results of partial regression analyses exploring how components of reproductive success in song sparrows depend on population density and the proportion of nests parasitized by cowbirds.

Component	Number of Females = P (Partial R^2)	Proportion of Nests Parasitized = P (partial R^2)	Nature of Relationship
Mean onset of laying	0.35 (0.03)	0.37 (0.03)	None
Mean end of laying	0.025 (0.20)	0.28 (0.04)	Earlier at higher densities; no effect of parasitism
Mean clutch size (sparrow eggs)	0.09 (0.07)	0.016 (0.17)	Smaller with more parasitism; no effect of density
Proportion of eggs hatching	0.50 (0.01)	0.001 (0.38)	Fewer with more parasitism; no effect of density
Proportion of nestlings fledging	0.91 (0.00)	0.62 (0.01)	None
Proportion of fledglings reaching independence	0.049 (0.15)	0.39 (0.03)	Lower at high densities, no effect of parasitism

Note. P = probability, R^2 = partial regression coefficient.

tion of nestlings that fledged. Finally, the proportion of fledglings that reached independence was greater at low densities, but it was independent of parasitism (table 5.2).[3]

In summary, population density and the proportion of nests parasitized both reduced ARS in song sparrows, but they did so by affecting different components of ARS. Density affected the end of laying and the survival of fledglings. Parasitism affected clutch size and the proportion of eggs that hatched.

Tests of the Cowbird Predation Hypothesis

The cowbird predation hypothesis makes four predictions. First, nest failure rates should be higher in years with cowbirds. Second, more nests should fail in years with greater reproductive activity by cowbirds. Third, nest failures should increase in late spring when migrant cowbirds arrive. Finally, parasitized nests should fail less often than unpar-

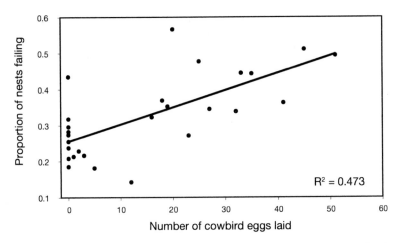

Figure 5.4. Proportions of song sparrow nests failing on Mandarte Island in relation to the number of brown-headed cowbird eggs laid per year.

asitized nests, because cowbirds should not interfere with nests containing their own eggs or nestlings.

First, as noted above, nests failed more often in years with cowbirds than in years without cowbirds (table 5.1).[4] Second, the number of cowbird eggs laid per year explained 43% of the variation in the proportion of nests that failed annually (figure 5.4).[5] Third, the timing of nest failures differed between years with and without cowbirds. In years without cowbirds, numbers of failures rose to a plateau in the first 30 days, where they remained for the next 70 days, before declining at the end of the breeding season. In years with cowbirds, numbers of failures rose steadily to a peak after 60 days, and 11–20 days after the onset of cowbird laying. They then declined from this peak level to the end of the laying season, but always remained above levels in years without cowbirds (figure 5.5).

Finally, a slightly higher overall proportion of parasitized song sparrow nests survived to hatching (76%, $n = 383$ nests) compared to unparasitized controls initiated at the same time in the season (69%, $n = 619$).[6] However, parasitism did not affect the proportion of sparrow nestlings that fledged (table 5.2). These results support the fourth prediction of the cowbird predation hypothesis only up to hatching. A nearby study (Rogers et al. 1997) found no differences in failure rates of parasitized and unparasitized song sparrow nests. That study, however, was conducted at a site where cowbird densities were high and interference between cowbirds was probably more common than on Mandarte Island.

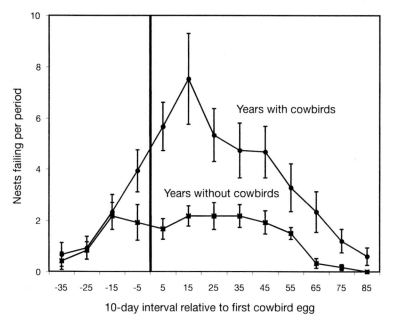

Figure 5.5. Mean numbers ($\pm$SE) of song sparrow nests failing in 10-day periods in 15 years with cowbirds (circles) and 12 years without cowbirds (crosses). The vertical line divides periods before and after the onset of laying by cowbirds in cowbird years. In years without cowbirds, the median date of laying by cowbirds in cowbird years was used to divide the data into "before" and "after" periods.

Variation in Nest Failure Type in Years with and without Cowbirds

To seek further evidence that nests failed for reasons associated with cowbirds, we used a set of standardized descriptions of nest failure (table 5.3). These types of failure differed with the presence or absence of cowbirds between 1982 and 2002.[7] Two types occurred only in cowbird years. Both of these fitted known cowbird activities: in type 3 failures, eggs were punctured or displaced from the nest cup, and in type 9, parasitized clutches were reduced in size and then abandoned. Together, these causes contributed 3.9% of added failures in years with cowbirds. Three rare types of failure, 6, 8, and 12, occurred in years with and without cowbirds but were statistically more common in years with cowbirds.[7] Two of these, type 8 (desertion after a reduction in clutch size) and type 12 (young dead in or near the nest) fit known

Table 5.3. Mean percentages ($\pm$SE) of song sparrow nests failing in 14 differing circumstances in 10 years with cowbirds and 12 years without cowbirds.

Circumstances of Nest Failure	Years Without Cowbirds	Years With Cowbirds	Chi-squared
1. Mouse droppings in empty nest; nest cup undisturbed	0.79 ± 0.35	4.00 ± 0.67	15.83^{***}
2. Broken egg shells; yolk on intact eggs	0.79 ± 0.35	1.06 ± 0.35	1.12
3. Punctured eggs or eggs outside nest	0	1.65 ± 0.44	—
4. Nest lining pulled out or disturbed	3.47 ± 0.73	7.29 ± 0.89	16.50^{***}
5. Nest overturned, destroyed, or missing	1.42 ± 0.47	1.41 ± 0.40	0.44
6. Vegetation near nest parted or trampled	0.32 ± 0.22	1.41 ± 0.40	5.50^{*}
7. Full clutch of eggs abandoned	3.94 ± 0.77	2.70 ± 0.56	0.07
8. Clutch abandoned after egg loss	1.58 ± 0.49	2.70 ± 0.56	4.77^{*}
9. Parasitized clutch reduced and abandoned	0	2.23 ± 0.51	—
10. Nest empty and undisturbed	5.52 ± 0.91	9.64 ± 1.01	16.3^{***}
11. Mouse droppings in nest + lining pulled	0.47 ± 0.27	0.82 ± 0.31	1.52
12. Damaged or dead young in nest or young outside nest	0.47 ± 0.27	2.00 ± 0.48	7.65^{**}
13. Young starved in nest; no sign of injury	3.00 ± 0.68	3.17 ± 0.60	1.36
14. Other	4.58	5.65	—
15. Success	73.66 ± 1.75	54.29 ± 1.71	—

Note. The chi-squared column reports whether the proportion of failures of a particular type differed between cowbird and no-cowbird years: $^{***}p < 0.001$, $^{**}p < 0.01$, $^{*}p < 0.05$. Probabilities could not be calculated for pairs with zero cases and are not presented for categories 14 and 15. See note 7 for more details.

patterns of cowbird behavior, but larger predators than cowbirds (e.g., river otters, glaucous-winged gulls) probably caused type 6 failures, which involved trampling and damage to vegetation at the nest.

Three common types of failure (1, 4, and 10) explained more than half of the increased failure frequency in cowbird years. Types 1 and 10 both involved nests that were found empty and undisturbed, but in type 1 failures mouse droppings were found in the nest cup. In type 4, the nest lining was disturbed. Did cowbirds contribute to these added failures? We suspect that they did from experimental work, where similar shifts in types 4 and 10 followed reduction in cowbird numbers (see below). Type 1 failures, however, may have arisen simply because the higher failure rates in cowbird years and the fact that mice often visited empty sparrow nests. We have previously tested, and failed to find support for, two plausible alternative hypotheses for increased nest failure in song sparrows at high densities on Mandarte: (1) predators such as deer mice and crows cause more nest failures in high-density years, and (2) cowbirds somehow facilitate losses to other nest predators (Arcese et al. 1996).

In summary, types of nest failure differed in years with and without cowbirds. Most, but not all, additional failures in cowbird years involved categories that matched the known consequences of cowbird behavior.

Experimental Tests of the Effects of Cowbirds on Song Sparrow Reproduction

We have shown above that cowbirds are associated with lower reproductive success of song sparrows on Mandarte Island, but it remains difficult to estimate the magnitudes of the reduction in ARS statistically because the age structure of the population, density, and the occurrence of cowbirds covary. Manipulative experiments provide a more stringent test cowbird effects, but when we removed cowbirds from Mandarte Island in 1977, we detected little difference in reproductive success or survival compared to four adjacent cowbird years (Smith 1981a).

We conducted a second removal experiment in the Fraser River Delta, about 50 km north of Mandarte, from 1995 through 1999. In this experiment, trapping reduced numbers of female cowbirds at two removal sites in 3 years (details in Smith et al. 2002). Parasitism at one of these sites was about four times as frequent (61%) as it is on Mandarte (15%; Rogers et al. 1997). The experimental reduction of cowbird numbers nearly doubled the ARS of song sparrows at both removal sites

Table 5.4. Contrasts between the annual reproductive success of female song sparrows at three study sites in relation to varying cowbird abundance.

Study Site	Fledglings per Female with No/few Cowbirds	Fledglings per Female with Cowbirds	Drop in Fledgling Production with Cowbirds
Mandarte Island	3.87 ± 0.42	2.77 ± 0.35	1.10
Westham Island	3.40 ± 0.34	1.86 ± 0.36	1.54
Delta Nature Reserve	3.16 ± 0.40	1.57 ± 0.29	1.59

Note. We compare mean fledgling production per female for 15 cowbird years and 12 years without cowbirds, after adjusting for effects of year and female age in Mandarte Island only. At the two other sites, values are numbers of sparrow fledglings produced in years when cowbird numbers were reduced and in pooled controls for the same years for the two other sites. Data for Westham Island and Delta Nature Reserve are from Smith et al. (2002).

(table 5.4; Smith et al. 2002). Reproduction improved because successful nests fledged more sparrows and failed less often (Smith et al. 2002), the same patterns seen between years with and without cowbirds on Mandarte Island. Three categories of nest failure were statistically more common in control treatments on the mainland: loss of entire clutches and broods, desertion of reduced and parasitized clutches, and attacks on nestlings (Smith et al. 2003). Again, these differences resemble those seen between years with and without cowbirds on Mandarte. It seems that many complete losses of clutches and broods in cowbird years are caused by cowbird predation, a conclusion that is supported by video monitoring (e.g., Granfors et al. 2001; Stake and Cimprich 2003).

Effects of Cowbird Parasitism on Adult and Juvenile Survival

The sight of a small songbird host feeding a large and voracious parasite chick immediately suggests that such a "mistake" might be costly for the foster parents (Hauber 2002). There have, however, been few analyses of this potential cost (Rothstein and Robinson 1998b). Similarly, few studies have quantified the effects of cowbird nest mates on juvenile survival after independence.

To test for costs to adult sparrows on Mandarte, we examined the survival of female and male sparrows that were parasitized, zero, one, or two or more times in cowbird years. We found that parasitism had no detectable effect on the survival of either males or females.[8] The power of this analysis is high; we could have detected a decrease in adult survival of only 6%. In a similar analysis, willow flycatchers also did not suffer reduced survival when parasitized by brown-headed cowbirds (Sedgwick and Iko 1999).

We made a final test for a survival cost of parasitism among juveniles. Independent juvenile song sparrows from broods that contained cowbird nestlings 5–7 days after hatching survived as well to breeding as sparrows without cowbird nest mates.[9]

Because raising cowbirds is energetically costly to song sparrows (Smith and Merkt 1980), the absence of survival costs here is surprising. It remains possible that the added cost of rearing a cowbird is paid through reduced fecundity in future (Gustafsson and Sutherland 1988), but we have yet to test this idea. In willow flycatchers, a much smaller host, a single parasite young often replaces a whole brood of host young (Sedgwick and Iko 1999). In species where parents do little additional work to rear cowbirds, the cost to adult survival may be small.

Defenses by Song Sparrows against the Cowbird

Although cowbirds frequently cause song sparrow nests to fail, we also know that song sparrows can defend their nests successfully against laying cowbirds. In our 1985 feeding experiment, only 7 of 38 nests (18%) of fed females were parasitized, compared to 42 of 94 control nests (45%). Supplemental feeding increased vigilance by males and incubation breaks by females, perhaps causing cowbirds to be detected and driven out more effectively (Arcese and Smith 1988). The strong dependence of parasitism on female age in song sparrows[1] also suggests that female attributes related to aggression and foraging efficiency may be vital to defense against cowbirds.

In summary, parasitism by cowbirds reduced reproductive success per nest and ARS per female sharply on Mandarte Island. Both cowbirds and high sparrow densities contributed to reduced ARS in sparrows on Mandarte Island, but density and cowbirds influenced different components of reproductive success. Cowbirds reduced clutch sizes and caused increased losses of sparrow eggs before hatching. However, cowbirds did not reduce the survival of adult song sparrows or juveniles reared in nests with a cowbird. Many of these results were confirmed by experiments wherein cowbirds numbers were reduced by trapping.

5.4. Discussion

The Song Sparrow and the Cowbird

Margaret Nice (1937) began the study of demographic interactions between avian brood parasites and their hosts. She documented several of the negative effects of cowbirds on sparrow reproduction described above. She also noted that cowbirds parasitized an increasing proportion (up to 78%) of nests as the song sparrow population declined. Nice, however, was properly cautious about using these data to support Herbert Friedmann's (1929) claim that cowbirds limit host numbers (p. 160 in Nice 1937). She did, however, recommend the cowbird as a promising subject for further study.

The work we describe above has extended Nice's findings, and others have also contributed useful insights into the interaction (e.g., Hauber 2000; Hauber and Russo. 2000; McLaren et al. 2003). Perhaps the most interesting finding arising from this work is that the "parasitic" effects of cowbirds on song sparrows (i.e., the net costs of rearing parasite instead of host young) are supplemented by a strong "predatory" effect, whereby cowbirds induce frequent nest failures by destroying clutches and killing or removing host nestlings from nests. Furthermore, the overall demographic effect of the cowbird on the song sparrow on Mandarte Island is to act as a density-dependent check on reproduction.

Although it could be argued that density-dependent reproductive suppression through brood parasitism is unique to Mandarte Island, experiments on the continental mainland (Smith et al. 2002, 2003), a study in New York (Hauber 2000), and comparative analyses (Arcese and Smith 1999) all show that cowbirds are closely associated with nest failure in song sparrow populations. Other reports of nest depredation in the hosts of brown-headed cowbirds (e.g., Arcese et al. 1996; Stake and Cimprich 2003) and shiny cowbirds (Nakamura and Cruz 2000) imply that our findings might be replicated in other, similar hosts. While the effects of generalist brood parasites on their hosts are becoming better known, this knowledge is less complete than our understanding of the evolutionary interactions between brood parasites and their hosts (e.g., Rothstein and Robinson 1998a; Davies 2000).

Temporal Shifts in Parasitism

We also found that cowbird reproduction declined over time. Cowbirds bred in only half as many years after 1989 than previously, and laid only one-fifth as many eggs. We offer three explanations for this temporal change. First, the lower average sparrow densities on the island after 1990 may simply have been insufficient to elicit a cowbird functional response to host density (figure 5.2; see also figure 4.1). Second, cowbird numbers in British Columbia were close to stable from 1975 through 1989 (−0.9% annual trend), but declined 6% annually from 1992 through 2003.[10] Finally, Mandarte Island may have become less suitable as a breeding site for cowbirds because cowbirds use exposed perches when searching for song sparrow nests (Saunders et al. 2003). In chapter 2, we documented the loss of many such perches on Mandarte over our study. Whatever the underlying mechanisms, fluctuating levels of brood parasitism are an additional example of how environmental variability affects bird numbers (see chapters 1, 11).

Management Implications from Studies of Song Sparrow Populations

Nest failures induced by cowbirds are of management interest, because cowbirds threaten some endangered songbird species in the United States (Rothstein and Cook 2000) and contribute to poor demographic performance in many other species (Robinson et al. 2000). While few song sparrow populations are of conservation concern (Chan and Arcese 2002), our experimental results suggest that reducing cowbird abundance in populations with frequent parasitism would lower nest failure rates and increase annual fecundity substantially. Such effects might generalize to other host species (Whitfield et al. 1999; but see Stutchbury 1997).

Reducing cowbirds may not stimulate host population growth if the host is limited by survival in the nonbreeding season (Sillett and Holmes 2002). Indeed, the reduced breeding activity of cowbirds after 1989 may not have allowed sparrow numbers on Mandarte Island to grow after 1995 because juvenile survival limited numbers overall (see chapter 4). However, cowbird reduction may "buy time" for threatened species with low reproductive rates. In 1971, the last population of the Kirtland's warbler numbered about 400 birds, and had declined more than 2-fold over a 10-year period (Rothstein and Cook 2000). After cowbird reduction began in 1972, fledgling production more than doubled and the population stopped declining for 15 years. After a fire created new breeding habitat, warbler numbers increased 6-fold (DeCapita 2000). In chapter 10, we model the effect of cowbirds on the size and extinction risk of the Mandarte Island population to explore their influence further.

Cowbirds as a Conservation Threat

Cowbirds first became widely perceived as a conservation threat in the 1980s, when Brittingham and Temple (1983) claimed that cowbird numbers were increasing rapidly. Terborgh (1989) used this claim and early results on population trends of migrant forest songbirds (Robbins et al. 1989) to label the cowbird as a conservation "villain" and the cause of population declines. These studies stimulated much new work on cowbirds and their hosts. Over the following decade, this research (summarized in Ortega 1998; Morrison et. al. 1999; Smith et al. 2000) improved our knowledge of cowbird–host relations greatly and also suggested different role for cowbirds in songbird declines.

Research in the 1990s revealed that cowbird numbers were stable from 1966 through 1976 (Sauer et al. 2005) but have declined thereafter. Although regional variation was high, cowbird and host numbers were generally positively correlated, opposite to the simple expectation if cowbirds were causing their hosts to decline (Peterjohn et al. 2000; Weidenfeld 2000). Furthermore, songbirds in other habitats, particularly grasslands, were declining faster than forest dwellers (Franzreb and Rosenburg 1997; Herkert et al. 2003). Thus, while cowbirds do contribute to poor performance of some songbird populations, their effects are only one of many landscape-level changes driven by humans (Robinson et al. 2000; Thompson et al. 2002).

Aggressive cowbird management is used to protect some endangered species from brown-headed cowbirds in the United States, with variable success (Rothstein and Cook 2000). Ortega (1998) argued that aggressive management might not be necessary, echoing Smith's (1994) comment that the cowbird is a convenient scapegoat for landscape damage caused by humans. Nevertheless, killing cowbirds remains a relatively cheap and effective management action, if cowbirds are parasitizing large fractions (e.g., >50%) of nests in an endangered host population and causing large reductions in fledging success in parasitized nests (Smith 1999). The shiny cowbird may be a greater conservation threat. This relative of the brown-headed cowbird threatens several of endangered hosts with extinction in the Caribbean region and South America (e.g., Woodworth 1999; Oppel et al. 2004).

5.5. Conclusion

Brood parasitic brown-headed cowbirds visited Mandarte Island more frequently at high sparrow population densities, laid more eggs, caused increased nest failure rates in sparrows, and facilitated density-dependent reproductive losses. Later in the study, cowbirds bred less often, and consequently had less impact on sparrow reproduction. Parasitism did not affect the survival of adult or juvenile song sparrows. Because cowbirds disrupt reproduction in several ways, and appear and disappear locally, they generate environmental stochasticity in the dynamics of Mandarte Island song sparrows. Cowbirds may have had similar effects on other hosts encountered during their spectacular range expansion over the past two centuries, but they are no longer increasing in abundance. Initial claims of damaging negative effects on their hosts may have been exaggerated.

Notes

1. Mean clutch size, number of fledglings, and number of independent young per nest were estimated using a General Linear Model (GLM), with year and female age included as categorical factors to account for their known influence on reproduction. Females 5 or more years old were pooled to enforce sampled sizes larger than five, and only the years with >5% parasitism were analyzed to control for the possibility that years and without parasitism differed in ways that might confound our results. Data were weighted by sqrt(attempts) to account for the fact that estimates of mean ARS per nest vary in precision. Parasitism reduced the mean number of fledglings per nest significantly ($p = 0.001$, $n = 455$) after accounting for effects of year ($p < 0.001$) and female age ($p = 0.16$). There was no significant interaction between age and parasitism ($p = 0.16$). Parasitism also reduced the mean number of independent young per nest ($p = 0.001$, $n = 455$) after accounting for effects of year ($p < 0.001$) and female age ($p = 0.06$). In this case, there was a significant interaction between age and parasitism ($p = 0.041$).

Because many females contributed to observations in more than one year and varied in the number of nests attempted annually, we also explored the effects of weighting estimates of mean female output by the square root of the number of nests attempted and the application of general linear mixed models (GLMM). Weighting improved estimates by reducing standard errors but had little effect on estimates of statistical significance. GLMMs, wherein female identity was treated as a random effect, gave very similar results to standard GLM, probably because intraclass correlation coefficients were generally low (<0.20) and a large fraction of females were included for only one year.

ARS was estimated as above. Females that were parasitized at least once reared fewer fledglings per year ($p = 0.002$) after accounting for effects of year ($p < 0.001$) and age ($p = 0.003$). Similarly, numbers of independent young were reduced by parasitism ($p = 0.002$) after accounting for effects of year ($p < 0.001$), and female age ($p = 0.002$). Interactions between age and parasitism were not significant for numbers of fledglings ($p = 0.10$), but they were significant for numbers of independent young ($p = 0.013$).

2. Regression analysis: Proportion of nests failing in years with cowbirds versus density (figure 5.3a), $b = +0.004$, $p = 0.03$, $R^2 = 0.31$; in years without cowbirds (figure 5.3b), $b = -0.001$, $p = 0.38$, $R^2 = 0.08$.

3. Results of linear regression models of effects of density and proportion of nests parasitized per year on timing and success of reproduction. (a) Overall model predicting numbers of independent sparrow young per female ($p < 0.001$, $R^2 = 0.49$); effects of density ($p = 0.003$, partial $R^2 = 0.24$), proportion of nests parasitized ($p = 0.04$, partial $R^2 = 0.10$). (b) Models of components of ARS. Mean onset of laying was independent of density ($p = 0.35$, partial $R^2 = 0.03$) and parasitism ($p = 0.37$, partial $R^2 = 0.03$); mean end of laying was later at high densities ($p = 0.025$, partial $R^2 = 0.20$) but independent of parasitism ($p = 0.28$, partial $R^2 = 0.04$); mean clutch size for sparrows was independent of density ($p = 0.09$, partial $R^2 = 0.07$) and lower with more parasitism ($p = 0.016$, partial $R^2 = 0.17$); mean proportion of sparrow eggs hatching was independent of density ($p = 0.50$, partial $R^2 = 0.01$) but lower in years with more parasitism ($p < 0.001$, partial $R^2 = 0.38$); mean proportion of hatched sparrows fledging was independent of density ($p = 0.91$, partial $R^2 = 0.00$) and parasitism ($p = 0.62$, partial $R^2 = 0.01$); and the mean proportion of sparrow fledglings reaching independence was lower at higher density ($p = 0.05$, partial $R^2 = 0.15$) but independent of parasitism ($p = 0.39$, partial $R^2 = 0.03$). $n = 27$ years in all models, except those involving numbers of fledglings, where n = 25, because numbers of fledglings were only known incompletely in 1975 and 1976. All experimentally fed females were removed from the data set and proportions were arcsine-square root transformed.

4. Nest failure rates in cowbird years with feeding experiments (1979, 1985, 1988) were calculated only for control pairs that did not have access to feeders. The proportions of nests failing in cowbird years and years without cowbirds differed (one-way ANOVA: $F = 4.905$; df = 11, 14; $p < 0.005$).

5. Regression analysis: arcsine-square root (probability of failing) versus number of cowbird eggs laid per year ($p < 0.001$, $R^2 = 0.37$).

6. Control nests here are those initiated from 7 days before the onset of cowbird laying to the end of the cowbird laying period each year, and control and parasitized nests are pooled across years: $G = 5.87$, df = 1, $p < 0.01$ (G-test with Williams correction, one-tailed).

7. Data in table 5.3 are for nests found at the egg stage only from 1982–2002. Nests of birds involved in feeding experiments in 1985 and 1988 were excluded. We analyzed these data with a multinomial logit model. It tested if overall proportions of failure types differed between cowbird and no-cowbird years ($\chi^2 = 56.5$, df = 1, $p < 0.001$) and whether the probabilities of a particular failure class differed from the probability of nests succeeding in cowbird versus no-cowbird years (for these probabilities, see table 5.3).

8. A generalized linear mixed model was used to analyze the probability of survival contingent on being parasitized none, once, or two or more times in cowbird years. We assumed a binary error distribution and a logistic link function. Individual bird was the random factor. We included year, age, sex, and the number of breeding attempts as covariates. While these covariates all influenced survival significantly, parasitism did not ($p = 0.34$). When we asked whether survival differed between birds that did or did not fledge cowbirds successfully, again, there was no difference.

9. We calculated the mean proportion of independent young surviving locally from 24 days of age to breeding age for (a) sparrow juveniles that had no cowbird nest mates at 5–7 days of age, versus (b) those that had one or more cowbird nest mates. Data are from all cowbird years, except 1979, as no data on juvenile survival were collected from 1979 through 1980. The respective proportions of juveniles surviving their first year were almost identical ($0.33 \pm$ SE $= 0.04$ without cowbirds, 0.32 ± 0.07 with cowbirds).

10. Cowbird numbers in British Columbia did not change significantly from 1975 through 1989 (trend $= -0.9\%$ per year, $p = 0.41$), but they declined significantly from 1990–2003 (trend -5.7%, $p = 0.001$). Data and linear route regression analysis from the Breeding Bird Survey (Sauer et al. 2005).

6 Social Mechanisms: Dominance, Territoriality, Song, and the Mating System

James N. M. Smith, Peter Arcese,
Kathleen D. O'Connor, and Jane M. Reid

In this sketch, a male floater (on the right) sings persistently while attempting to usurp a territory from its owner in spring. The owner approaches silently and threateningly. Typically, such contests result in the owner driving off the floater, but some end with the owner losing part or all of its territory.

Social dominance and territoriality are related aspects of aggression (Watson and Moss 1970). Dominance occurs when an individual repeatedly gets first access to a benefit such as food or a social opportunity (Piper 1997). Dominance is often signaled by threats from dominants and appeasement or submissive behavior by subordinates. Territoriality occurs when dominants exclude conspecifics from a particular area (Brown 1969). Territoriality and dominance thus differ mainly in the degree to which an individual monopolizes space. Often, only some parts of a joint range are defended while others are shared amicably and rank within a pair may shift across the range (Brown 1964). Both dominance and territoriality are conspicuous in songbirds. Territoriality is widespread in the breeding season (Howard 1920), while dominance occurs in many nonbreeding flocks (e.g., Nolan and Ketterson 1990).

The idea that the aggressive behavior of individuals limits animal numbers has a long history (Nice 1943). Bernard Altum (1868) first recognized that male bird song threatens male rivals and attracts females. Eliot Howard extended Altum's ideas in his influential book (Howard 1920, p. 272), where he suggested that territoriality limits breeding numbers. Stewart and Aldrich (1951) and Hensley and Cope (1951) later tested Howard's idea experimentally. They removed territorial breeders from several populations of forest songbirds and showed that replacement individuals soon filled the vacancies. Further studies (reviewed in Newton 1992, 1998) confirmed these early findings and showed that territorial populations often contain many nonterritorial "floaters" (e.g., S. M. Smith 1978). In rufous-collared sparrows, floaters live cryptically within ranges overlapping a few contiguous territories (S. M. Smith 1978). Floating horned owls share communal areas between territories that are lightly used by owners (Rohner 2004). In the woodland-dwelling great tit, some floaters disperse to breed in less productive hedgerow habitats (Krebs 1971). Nonterritorial red grouse move to habitats where death rates are high (Watson 1985).

Dominance also influences the survival and breeding status of individuals. In the complex societies of some tropical birds (e.g., McDonald 1993) and mammals (Alberts et al. 2003), dominance "queues" may form for breeding opportunities. In the simpler social systems of temperate songbirds, age and gender groups may differ in dominance (e.g., Nolan and Ketterson 1990) and survival. Subordinates are sometimes excluded from the best feeding sites or habitats (e.g., Drent 1984;

Marra 2000). Population growth rate in birds often depend critically on juvenile survival in winter (Newton 1998). Hence, intraspecific competition in winter should promote increased mortality (Lomnicki 1980; Kikkawa 1980) or dispersal (Ekman et al. 2002) among subordinates.

Song is the principal means used to proclaim territory ownership by male songbirds (Kroodsma 1996). Females may also sing (Richison 1983) particularly in species that are territorial year round (Arcese et al. 1988; Bard et al. 2002). Songs are learned either from an individual's father or from neighbors when a young bird establishes a territory for the first time (reviewed in Kroodsma 1996). In birds with complex song repertoires, a large repertoire may confer high mating success and fitness (e.g., Catchpole and Slater 1995; see below).

The mating systems of birds have long fascinated ornithologists (e.g., Gilliard 1969). Two themes have dominated the study of avian mating systems: first, the origin and maintenance of the three principal types (monogamy, polygyny, and polyandry; Emlen and Oring 1977; Searcy and Yasukawa 1995; Davies 1992), and second, sexual selection (Andersson 1994). Mating systems, however, also affect population structure and life history tactics in several ways (Shuster and Wade 2003).

The total population size, N, is the number of individuals within a population. However, a more important quantity for assessing the conservation status of small and threatened populations is the *genetically effective population size*, N_e, the number of individuals that actually breed and contribute to future generations. Social dominance and the mating system affect N_e, because subordinate individuals may not pass genes on to future generations. In "harem" polygyny, a few males may sire most young. As a result, N_e can be much smaller than N (e.g., Hoelzel et al. 1999). In monogamous systems, N_e might be expected to equal N. However, extrapair mating is common in some apparently monogamous species (e.g., Westneat 1987; Birkhead and Møller 1992). If a few successful extrapair sires monopolized such extrapair matings N_e could be reduced from that expected given social monogamy (Waite and Parker 1997).

In this chapter, we first describe how dominance and territoriality contribute to the limitation of numbers in the Mandarte Island song sparrows. Next, we consider territory acquisition and mate fidelity and its reproductive consequences. We then investigate whether song repertoire size indicates male fitness in Mandarte song sparrows. Finally, we describe the social and genetic mating systems of the song sparrow on

Mandarte and ask whether extrapair mating reduces the effective population size.

6.1. Dominance

Aggression in song sparrows develops among fledglings before they leave their natal territory (Nice 1943). On Mandarte Island, newly independent juveniles gather each summer in flocks where perches and feeding sites are numerous. In these flocks, dominant juveniles threaten rivals with body feathers raised, crest flattened, the wings held out from the body, and beak aimed at a rival. They often give *zhee* threat notes. Subordinates stand erect with feathers sleeked, crest raised, and wings flattened against the body. They frequently give *tsip* calls and retreat when a dominant approaches (Nice 1943). By late summer, early-hatched young generally dominate late-hatched individuals. Dominant young of both sexes in these flocks are more likely to survive locally and obtain breeding territories on Mandarte than are subordinates (Arcese and Smith 1985). Juvenile females that disperse from the island also tend to be subordinates (Arcese 1989a).

Dominance interactions remain conspicuous throughout winter, particularly at concentrated food sources. Adult males generally dominate yearling males and adult females in winter, and all of these classes dominate juvenile females (Knapton 1976; Smith et al. 1980). Work on other North American sparrows (reviewed by Piper 1997) has shown that dominance in winter depends on a variety of factors including: sex, age, hunger, plumage, and local residence time. If winter food limits population growth on Mandarte Island, we might expect juvenile females to suffer the highest mortality, or to disperse most frequently, because of their low rank at feeding sites (Smith et al. 1980). We showed in chapter 4 that males outnumbered females substantially in some yearling cohorts and that adult females also did not survive as well in these years. We therefore speculate that the low dominance status of adult and, particularly, juvenile females during winter were responsible for poor local female survival in these years.

6.2. Territoriality

Territory Acquisition by Males

In resident song sparrow populations, males compete conspicuously for territories with neighbors and nonterritorial birds over much of the year

(Knapton and Krebs 1974; Wingfield and Soma 2002). Territorial contests are particularly noticeable in early spring, when neighboring pairs display along a common territorial boundary (Nice 1937).

The use of territories varies across the species' range. In Ohio, all foraging and reproductive activities are conducted on territories (Nice 1941). On Mandarte, territorial birds often leave their territories to bathe or to forage in the intertidal zone (see chapter 11, title sketch) and undefended grass meadows. In winter, territory holders join feeding flocks of nonterritorial sparrows for part of the day (Arcese 1989b).

On Mandarte, changes in territory ownership occurred throughout the year, but turnovers peaked in fall and spring (Arcese 1987, 1989a). Typically, yearling floaters settled between defended territories, sang persistently (title sketch, this chapter), and took over space from one or both of the previous owners; 35% of 292 observed turnovers between 1982 and 1987 followed this pattern (Arcese 1989b). Owners typically responded by driving the intruder from the territory immediately, but some contests lasted up to 2 days. In these prolonged contests, and when more than one intruder participated, owners lost much or all of their territories (Arcese 1987). About 25% of deposed territory owners became floaters on their previous territory, and some of these individuals eventually regained a nearby territory (Arcese 1989b). Male neighbors accounted for 30% of 292 territory takeovers and sometimes also gained a second mate by this means. Age influenced the outcomes of contests for territories. Males 2 and 3 years old were more likely to win contests with neighbors than were yearlings or males 4 years or older (Arcese 1989b).

Territoriality, Population Density, and Mated Status in Males

On Mandarte Island, territorial males defended almost the entire shrub area regardless of population density (figure 6.1). Because population size fluctuated, there was a strong negative relationship between territory size and the number of males with territories each spring (figure 6.2). The larger territories of unpaired males were generally used for breeding in other years (compare figure 6.1, a vs. b). However, some yearling males defended small patches of scrub (<100 m^2) that were seldom used for breeding. The territories of unmated males were smaller on average than those held by monogamous males (figure 6.2). This pattern varied with male age in a complex way. Most variation in territory size with age was within mated males. Territories were smallest in mated yearlings, increased sharply to age 2, and then declined to age

a) 1985

b) 1989

Figure 6.1. Two extreme territory arrangements for 1985 (75 territorial males) and 1989 (7 territorial males). Solid areas represent shrubs that were defended by paired males; shaded areas represent shrubs defended by unpaired males. Open outlined areas within the shrub margins were not defended by males.

4. The few very old males had relatively large territories (figure 6.3). Population size and a male's mated status and age accounted for 31% of the variation in territory size over the study.

Recruitment in male song sparrows can be stimulated experimentally by creating several simultaneous territorial vacancies (Knapton and Krebs 1974). However, episodes of settlement after catastrophic losses did not influence male recruitment on Mandarte Island between 1975 and 2003. Males established most new territories on Mandarte sequentially during the nonbreeding season (Arcese 1989b), rather than simultaneously after catastrophic mortality episodes. There was, however, one exception to this pattern after a spring snowstorm in 1962 (Tompa 1971).

On average, 20% of territorial males were unmated, but this percentage increased sharply as the sex ratio became more male biased after 1992. Up to 1992, 12% of males were unmated, but the average rose to 33% after 1992, with a peak of 65% in 1999 (figure 6.4).

Territory and Mate Fidelity

Once a male gained a territory in the main area of shrub on Mandarte, he was likely to retain it if he survived. Males that owned territories in

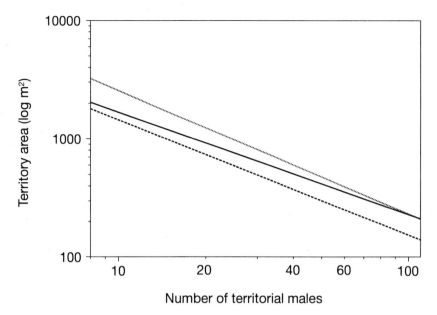

Figure 6.2. Areas defended by male song sparrows at varying male densities. Dotted line, socially polygamous males; solid line, monogamous males; dashed line, unmated males. Log_{10} territory area declined as the number of territorial males increased (adjusted $r^2 = 0.25$, $F_{1,1217} = 421.33$, $p < 0.001$). Adjusted least-squares means (square mean $\pm$ SE) for socially polygamous, monogamous, and unmated males were 417 (1.0), 367 (1.0) and 269 (1.0) m^2, respectively. Overall, type III general linear model with male mated status included as a categorical variable: $F_{2,1217} = 35.78$; $p < 0.001$.

successive years rarely moved, other than through minor adjustments to their territory boundaries. A full analysis of mate fidelity and its consequences for the reproductive performance of males and females on Mandarte has yet to be done. However, we illustrate patterns of territory and mate fidelity here with data from 1986 through 1988, years of similar and above-average density. In these years, all but two territorial males had a mate in each year. From 1986 through 1988, 97% of 80 surviving males retained their territories from one year to the next. Two males moved one and two territories distant from the territory they occupied the previous year.

Strong fidelity of individuals to territories was matched by fidelity to mates. From 1986–1988, 48 of 54 surviving pairs (89%) remained

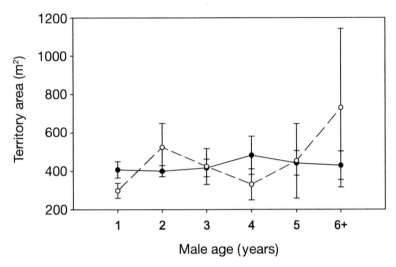

Figure 6.3. Mean territory area defended by song sparrow males of different ages. Open circles represent mated males; solid circles represent unmated males. Log_{10} territory area varied with male age, with territory size increasing from age 1 to age 2 but remaining constant thereafter. General linear model with number of males included as a covariate and social status as a categorical predictor ($F_{5,1105} = 8.27$, $p < 0.001$).

together from one year to the next. All females that paired to the same male retained the same territory across years. However, when surviving pairs broke up, females moved an average of five territories away (SE = 1.8, $n = 6$). When a female's previous mate died or lost his territory, females ware about equally likely to retain the territory (12 of 26 cases) or to move (14 cases). The latter females moved an average of 2.1 territories (SE = 0.7, $n = 14$).

Because song sparrows rear multiple broods per year, females can switch mates within a year; on average 15% of females did so (SE = 0.02%, $n = 27$ years). In most of these mate switches, the female remained on the same territory after her male died or was displaced by a rival. Mate switching was most common from 1992–2001, when breeding sex ratios became strongly male biased. Mate switches in mid-season had no immediate cost to females. In fact, females that changed mates within a season reared slightly more young per year than did females that remained with the same mate.[3]

In most other short-lived birds, mate fidelity is generally lower than in the song sparrow (Dhondt et al. 1996; Désrochers and Magrath 1996;

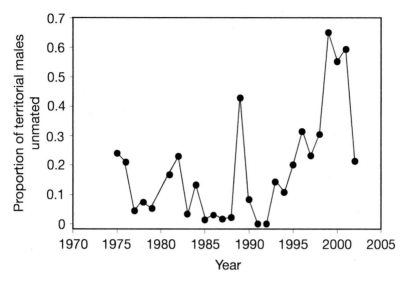

Figure 6.4. Variation in the proportion of song sparrow males that were unmated during the study.

Payne and Payne 1996; but see Murphy 1996). Individuals that remate generally increase their reproductive success by doing so (Ens et al. 1996).

Effects of Density on Male Floaters

If territoriality acts strongly to limit male numbers, we expect both the number of floaters, and the proportion of the male population that becomes a floater, to increase with density. The number of floating males was indeed higher at high densities ($r = 0.45$), and the fraction of older male floaters also increased with male density ($r = 0.32$).[2]

Reproductive Success in Male Floaters

Most early studies (reviewed in Watson and Moss 1970) considered floaters to be poor-quality individuals with low reproductive potential. During the 1980s, we studied the social strategies and reproductive success of floaters on Mandarte in detail (Arcese 1987, 1989a; Smith and Arcese 1989) and found that many former floaters gained territories after their onset of their first breeding season (Arcese 1989b). Males that were floaters during their first year produced fewer offspring over their lifetimes than did males that mated as yearlings, but they preformed as well over their lives as yearling males that had a territory but no mate (Smith and Arcese 1989).

Territoriality and Dominance in Females

Female song sparrows fight less conspicuously than males, but threats between females are common in spring and may escalate to singing and fighting at high population densities (Nice 1943; Arcese et al. 1988). For example, in April 1985, a female floater attacked an incubating female on Mandarte at her nest. The incubating female's nest was subsequently destroyed, the rival took over the territory, and the original owner was not seen again (Arcese 1989c). Female floaters, however, almost never occurred after 30th April. Perhaps, female–female aggression near the onset of breeding forced female floaters to disperse. Indeed, over the entire study, several unbanded females arrived on Mandarte Island just after the onset of breeding in April and early May and immediately bred. Once nesting began, females displayed marked aggression toward each other only when two or three females lived with a male in polygynous groups. Such females occasionally fought and often interacted aggressively. Female fighting is also common near the onset of breeding in blue tits (Kempenaers 1995).

Female settlement and territoriality have yet to be studied in detail in song sparrows on Mandarte. However, periods of emigration by juveniles in autumn and late winter occur at about the same time as peaks in territorial activity among adults. This coincidence suggests that territorial interactions may influence the rate of female dispersal from the island in autumn and late winter (Arcese 1989a). The factors influencing female recruitment in other songbird species also remain poorly understood.

In summary, patterns of aggression among adults, and disappearance of young birds of each sex from Mandarte Island, suggest that population size is influenced by social behavior in both fall and spring (Tompa 1971; Arcese 1989b; Arcese et al. 1992). Indeed, the fraction of the juvenile cohort that bred as a yearling declined markedly as the number of territorial adult females in the breeding period increased (see figure 4.2). Males were much more likely than females to remain on the island as nonbreeding yearlings.

6.3. Song

Territoriality in song sparrows is accompanied by loud advertising song given from the territory by an owner. Males learn up to 16, but commonly 6 to 10, distinct song types each made up of four or five phrases (Mulligan 1966; Podos et al. 1992; Beecher et al. 2000; Reid et al. 2004). Song sparrows apparently treat song types as natural units that transcend finer scale variation in song structure (Searcy et al. 1999). Intruding birds also sing persistently when attempting to take over territories from owners (Arcese 1987, 1989a; Bower 2000). Western song sparrows learn most of their songs from territorial neighbors in their natal neighborhood over their first nine months of life (Cassidy 1993; Hill et al. 1999; Nordby et al. 1999; Wilson et al. 2000). As a result of this learning process, male neighbors share many identical songs, and they retain these songs for the rest of their lives (Cassidy 1993; Nordby et al. 1999; Wilson et al. 2000; Searcy et al. 2002).

Song and Reproductive Success in Males

Song diversity is a classical example of a sexually selected trait (reviewed in Searcy and Yasukawa 1996), and might therefore be expected to predict a male's reproductive success. Song sparrows have been well studied in this regard, but results have not been consistent across studies. In laboratory experiments, females displayed more to playback of songs from males with larger repertoires of songs, suggesting that females prefer males with larger repertoires (Searcy and Marler 1981; Searcy 1984). In contrast, free-living males with larger repertoires did not pair earlier, or obtain replacement females sooner, after their original mate was removed experimentally (Searcy 1984). Hiebert et al. (1989) found that on Mandarte Island, male song sparrows with larger repertoires acquired territories earlier, held them for longer, and raised more offspring to independence. However, two other studies failed to find positive correlations between repertoire size and male phenotype or the length of time a male held a territory (Searcy et al. 1985; Beecher et al. 2000).

To try to resolve these contradictions, we recently revisited the relationships between male song repertoire size, mating success, and lifetime reproductive success on Mandarte (Reid et al. 2004, 2005a, 2005b). We found that yearling males that sang more song types were more likely to gain mates by their first April (figure 6.5a), and that females mated to yearling males with larger repertoires bred earlier in the year (figure 6.5b). Males with larger song repertoires also lived longer and left more first-generation (figure 6.5c) and second-generation descendents (figure 6.5d) than did males with smaller repertoires.

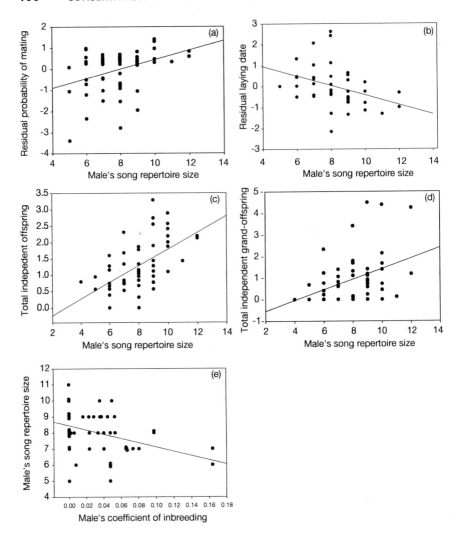

Finally, we investigated whether a male's song repertoire size indicated his inbreeding level, a measure of his relative level of genetic heterozygosity (and therefore genetic diversity). Indeed, we found that inbred males sang fewer song types than did outbred males (Reid et al. 2005a; figure 6.5e). We additionally found that males with small song repertoires mounted a weaker cell-mediated response to an experimental immune challenge (Reid et al. 2005b). In conclusion, these results and laboratory studies showing that female song sparrows are more stimulated by playbacks of larger repertoires (e.g., Searcy 1984) suggest that song repertoire size is a sexually selected trait in song spar-

Figure 6.5. Relationships between a male song sparrow's song repertoire size (his total number of different song types) and (a) the probability that a yearling male would mate by his first April, (b) the breeding time of a yearling male, (c) the total number of first-generation offspring produced, (d) the total number of second-generation offspring produced and (e) the male's coefficient of inbreeding. The probability that a yearling male would mate increased with his song repertoire size, after controlling for population sex ratio and the size of the male's territory (a; logistic regression, sex ratio $\chi^2_1 = 4.4$, $P = 0.035$, repertoire size $\chi^2_1 = 5.2$, $P = 0.023$, year-standardized territory size $\chi^2_1 = 9.8$, $P = 0.002$). Females mated to yearling males with larger repertoires laid their first egg earlier in the season after controlling for between-year variation in laying date and the size of the male's territory (b; multiple regression, year mean lay date $F_{1,41} = 22.2$, $P < 0.001$, repertoire size $F_{1,41} = 6.4$, $P = 0.016$, territory size $P = 0.37$). After standardizing each male's observed lifetime reproductive success for his cohort mean, males with larger repertoires left more offspring and grand-offspring on Mandarte than did males with small repertoires ($r = 0.57$, $P < 0.001$ and $r = 0.38$, $P = 0.005$ respectively). More inbred males sang smaller repertoires (e; $F_{1,50} = 15.9$, $p < 0.001$).

◄──────────────────────

rows. Females may prefer males with larger repertoires because repertoire size acts as a reliable indicator of a male's genetic quality, immunity, and lifetime reproductive success.

Other studies of repertoire size, male survival, and lifespan in song sparrows, however, have shown that it may not necessarily be repertoire size per se that is linked to male fitness. The number of songs that a male shares with his neighbors in Seattle (Beecher et al. 2000) and the distance a male disperses in the San Diego area (Wilson et al. 2000) are also correlated with his reproductive success. Repertoire size and numbers of songs shared with neighbors are positively correlated, at least up to a repertoire size of nine (Beecher et al. 2000). Further work is needed to identify the precise target of sexual selection in different populations of song sparrows. There is recent evidence of spatial and temporal variation in sexual selection acting on repertoire size in great reed warblers (Forstmeier and Leisler 2004).

In summary, fieldwork on Mandarte Island and laboratory work have shown that song repertoire size predicts mating success of male song sparrows early in life and that females prefer males with larger repertoires. In addition, repertoire size and the degree to which males match their neighbors' songs predict lifetime fitness well in male song sparrows.

6.4. The Social Mating System

Like many other small songbirds (Elphick et al. 2001), song sparrows on Mandarte have a monogamous mating system where most breeders (95.5%) live in pairs. Some birds, however, breed in polygynous groups. We observed 44 trios of a male and two females and five groups of one male and three females during 27 years of study, compared to 1,098 monogamous pairs. This frequency of polygyny (4.5%) approaches the traditional cutoff point (5%) for separating monogamous from polygynous bird species (Hasselquist and Sherman 2001). Polygynous groups formed at two times: before breeding began, and when a paired male died or lost his territory during the breeding season and a mated neighbor expanded to occupy his territory.

The predominance of monogamy in songbirds is usually explained by the hypothesis that polygyny reduces female fitness when two females have to share a male's parental contribution (Wittenberger and Tilson 1980). Frequent polygyny in songbirds can occur when territory and/or female quality vary, so female reproductive success is higher in a trio on a good territory than in a pair on a poor one (Forstmeier et al. 2001). Polygyny can also occur when there is a local imbalance of the sex ratio with an excess of females (Emlen and Oring 1977).

In 1979, we manipulated the local sex ratio on Mandarte to test Emlen and Oring's (1977) sex ratio hypothesis. We temporarily moved nine males from their breeding territories soon after their first broods hatched. The removals caused three of the nine "widowed" females to form trios with mated neighbors, and one to form a group of four with a neighbor that already had two social mates. Three of the other females moved to mate monogamously with unmated or recently widowed males, one died before remating, and the ninth retained her territory but did not breed again (Smith et al. 1982).

We also explored the determinants of polygyny in our full 27-year data set. We initially excluded birds involved in the male removal experiment above and in three feeding experiments. As expected from our experiment, polygyny was more frequent when the population sex ratio was less male biased. There were six trios and two groups of four in 1988, the only year of 27 with more females than territorial males. The frequency of polygyny also depended on age, with nearly opposite patterns in males and females (figure 6.6). In females, polygyny was most frequent in birds 5 or more years old and least frequent at age 2. In males, polygyny peaked at age 2 and was least common at age 1 (fig-

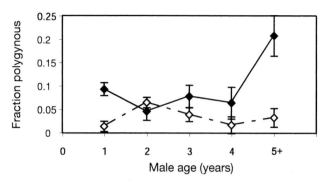

Figure 6.6. Frequency of polygyny in song sparrows of different age and sex (solid symbols, females; open symbols, males). Error bars = ±1 SE.

ure 6.6). However, male age was not a reliable predictor of polygyny when assessed on its own.[3] In a previous analysis of data up to 1986, Arcese (1989c) noted a similar pattern of age dependence for females. The pattern of age dependence for males was strong up to 1986, but that relationship weakened when new data were added. When we incorporated the male removal experiments into the logistic regression model, removals increased the probability of polygyny in females.[3]

Following Arcese (1989c), we interpret age-related variation in polygyny among females as a result of age differences in competitive ability. Since polygyny is costly to females (see below), highly competitive 2-year-old females are best able to avoid polygyny, while old females are the least able. These effects of age on polygyny also resemble effects of age on reproductive success (see chapter 3) where female performance declines with advancing age, and the effects of age on brood parasitism (Smith and Arcese 1994) where the oldest females are parasitized most often. Declining competitiveness in old age among females could explain all three results. Age-related differences in competitiveness also exist in male song sparrows (e.g., figure 6.3), but are strongest between yearlings and older birds.

There are good reasons to expect polygyny to increase in frequency when food for nestlings is not limiting (Verner and Willson 1969; Searcy and Yasukawa 1995). We supplemented food of song sparrows experimentally in spring in 1979, 1985, and 1988. When we included supplemental feeding as a covariate in the logistic regression analysis, the predicted frequency of polygyny rose from 10% to 19% in females, but it did not change in males.[3] The increased frequency of polygyny after supplemental feeding was not caused by new females settling on fed territories, as predicted by theory (Emlen and Oring 1977). Instead, fed males expanded their territory boundaries to encompass the ranges of additional females (Arcese 1989a, 1989c), a result also found in dunnocks (Davies and Lundberg 1984).

Costs and Benefits of Polygyny

Polygynous males reared more fledglings on average than monogamous males although the benefits of polygyny varied from year to year (figure 6.7). Thirty-eight males with more than one female reared an average of 0.72 additional young per year than did monogamous males breeding in the same years. However, there is a possible difficulty here. Females in polygynous groups might participate in more extrapair matings, overriding the apparent advantage of polygyny to males. Kempenaers et al. (1995), however, did not find this result in blue tits. Also, extrapair paternity in songbirds is generally less frequent in species with regular polygyny (Hasselquist and Sherman 2001).

We have data on paternity for five polygynous males from 1993 through 1996 (O'Connor 2003; see section 6.5). These males did not sire of 18% of young in their nests, compared to 28% for the remaining monogamous males. Therefore, while more data are needed, we have no reason to believe that song sparrow males in polygynous groups suffered high levels of extrapair paternity. As a consequence, polygyny is likely to have increased the fitness of male song sparrows on Mandarte.

In contrast, polygyny was costly to the average female. Females that shared a social mate for the entire season reared fewer young than females that spent only part of the year in a trio or group of four. Females that lived in groups for the entire season reared 1.31 ± 0.30 fewer fledglings and 0.88 ± 0.28 fewer independent young than did monogamously paired females breeding in the same years. Females that spent only part of the season in breeding groups suffered smaller losses, rearing 0.49 ± 0.30 fledglings and 0.53 ± 0.32 fewer independent young, respectively, than did monogamous females.[4] The estimated cost of season-long mate sharing to a female, 35% of her median annual reproductive output, is substantial.

It can, however, still be argued that the females that breed in groups maximize their fitness by doing so (Grønstøl et al. 2003), for example if the alternative is to mate monogamously with an unmated male on a poor-quality territory (figure 6.2). Further, in years such as 1988, no unmated territorial males were available, and females faced the choice of not breeding at all or breeding in groups. In other years, however,

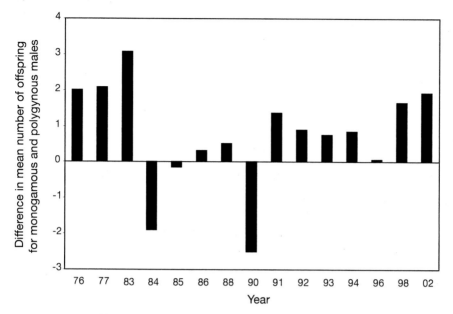

Figure 6.7. Difference between the mean number of independent offspring reared by polygynous and socially monogamous song sparrow males in years with polygyny. Bars above zero indicate that polygynous males reared more offspring than did monogamous males on average. Both years with greater productivity in monogamous males had only one polygynous male.

unmated males were available, and some widowed females did pair with these males. Territory and mate switching were not costly to such females (see section 6.2).

Many studies of polygyny distinguish "primary" and "secondary" females based on their order of settling. In general, secondary females pay a higher cost of polygyny (Grønstøl et al. 2003). We often did not know the order in which female song sparrows settled and thus could not distinguish primary from secondary females. However, a related mechanism does affect the costs of polygyny for female song sparrows. Males in breeding groups usually bring food only to the first brood to hatch. As a result, the offspring of unaided females grow more slowly and suffer higher mortality (Smith et al. 1982; Arcese 1989c).

In summary, this synopsis confirms several of the conclusions reached by earlier studies of the mating system on Mandarte Island (Smith et al. 1982; Arcese 1989c). Polygyny occurs regularly and is more common in years with a less male-biased sex ratio. Supplemental feeding raises the frequency of polygyny in females. Polygyny depends strongly on age in females, with 2-year-old females experiencing polygyny least often. In males, there is a different but statistically weaker pattern of age dependence. Polygyny is generally beneficial to males and costly to the average female, although it still may be the best option available to any individual female.

6.5. The Genetic Mating System

The social arrangements of animals are diverse and well described (e.g., Wilson 1975; Shuster and Wade 2003). In birds, most individuals live in seemingly monogamous pairs (Black 1996). When molecular parentage analysis was first applied to songbirds (Westneat 1987; reviewed in Birkhead and Møller 1992), ornithologists were astounded to find that extrapair parentage was common. It is therefore useful to distinguish between *social mates* and *genetic mates*.

Song sparrows resemble other songbirds in practicing social monogamy, with occasional social polygyny (see above). When we used nine microsatellite loci to assign parentage on Mandarte Island, we found that social fathers sired only 72% of all song sparrow chicks from 1993 through 1996 (O'Connor 2003). This value is typical for passerines as a whole, but high for island populations, where extrapair paternity is generally less frequent (Griffith 2000; Robertson et al. 2001).

When there are frequent extrapair fertilizations in a population, estimates of the effective population size (N_e) based on the social mating system may not be accurate (Waite and Parker 1997). We used Nunney and Elam's (1994) model to calculate N_e. This model incorporates the effects of sex ratio (r), adult life span (A_i), generation time (T), standardized variance in reproductive success (I_{bi}), and the standardized variance in life span (I_{Ai}):

$$N_e/N = \frac{4r(1 - r)T}{r\,(A_f\,(1 + I_{af}) + I_{bf}) + (1 - r)\,(A_m\,(1 + I_{am}) + I_{bm})}$$

A high fraction of extrapair fertilizations could lower the effective population size (N_e) if some paired males sire many more or many fewer offspring than fledge from their nests. Alternatively, some unpaired males might sire many offspring in the territories of mated males. In the Mandarte population, however, neither of these conditions was met; the extrapair sires were little more successful than the average territory owner (figure 6.8). From 1993 through 1996, most of the extrapair

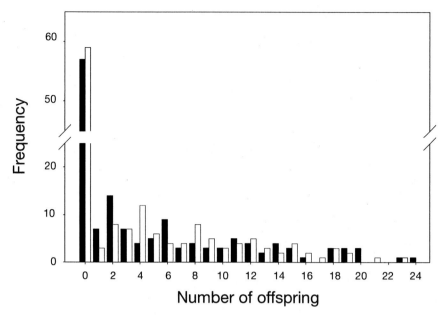

Figure 6.8. Histogram illustrating the number of individual males that produced a certain number of offspring during 1993 through 1996. Solid and open bars indicate the number of males that produced a certain number of offspring based on social and genetic information respectively.

sires were territorial neighbors within 50 m of the focal territory (figure 6.9). Mated males sired 85% of these extrapair young (see table 6.1), and unmated territorial males 14%; only one of 212 extrapair young was sired by a floater. Therefore, the effective population size that we estimated based on the social mating system did not differ from the estimate based on the genetic mating system by more than one individual in these 4 years.

During 1993–1996, effective population size averaged 64% of actual population size, and social N_e was a very good approximation of genetic N_e (O'Connor et al. in press). N_e is only to be expected to be lower than N in monogamous mating systems if a few individuals monopolize most extrapair matings (Waite and Parker 1997).

6.6. Conclusions

Territoriality in the song sparrow on Mandarte Island broadly resembles that described in resident populations of great tits (Drent 1984) and

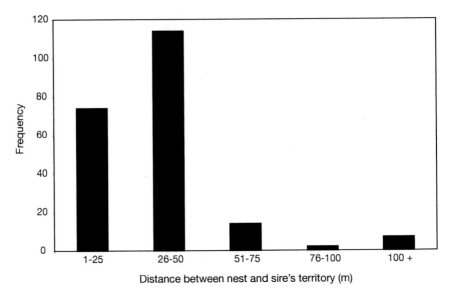

Figure 6.9. Distances between the nests of extrapair young that survived to at least day 6 and the center of the territory of their genetic sires.

Steller's jays (Brown 1964). Territorial individuals maintain year-round dominance over an area and exclude rivals more vigorously in the breeding season, but they do not exclude other conspecifics completely.

Two social factors influence recruitment of young song sparrows to the Mandarte Island population. First, competition for breeding territories causes many yearlings to die or disperse from the population during their first winter. The timing of these losses in Mandarte song sparrows differs from that seen in continental songbirds. In great and coal

Table 6.1. Mated status of male song sparrows on Mandarte Island from 1993 through 1996, and the mated status of the sire of all extrapair young (EPY) that survived to at least 6 days after hatching

Year	Number of Males	Number of Mated Males	Number of Unmated Males	Number of EPY	Number of sires Mated	Number of sires Unmated
1993	71	48	23	47	45	2
1994	75	58	17	44	43	1
1995	71	48	23	64	53	14
1996	82	48	34	54	40	14

tits (Dhondt 1979; Naef-Daenzer et al. 2001) and white-crowned sparrows (Petrinovich et al. 1981), most juvenile dispersal takes place during the first summer. The factors limiting local survival of juvenile females on Mandarte operate earlier in winter than those affecting males, because virtually no nonterritorial females remain on the island after April 30 each year.

Second, dominance status in late summer predicts survival in juvenile song sparrows on Mandarte Island. Dominant juveniles are much more likely to survive winter than subordinates (Arcese and Smith 1985), and subordinate females are more likely to disperse from the island (Arcese 1989a). Juvenile females are subordinate in winter (Smith et al. 1980), and high mortality among young females in less favorable winters may generate biased adult sex ratios (see chapter 4).

Song sparrows exhibit a social system, social monogamy with some polygyny, that is widespread among songbirds. Polygyny is more common in years with a female-biased sex ratio. When the sex ratio is more male biased, polygyny is rare, although females switch mates commonly. The oldest females are most likely to join breeding groups. Polygynous males generally rear more offspring than do monogamous males, and the opposite result applies to females.

Individual variation in a social trait, the numbers of songs in a male's repertoire, is a good predictor of a male song sparrow's lifetime reproductive success. Similar patterns have been found in great reed warblers (Hasselquist et al. 1996; but see Forstmeier and Leisler 2004) and great tits (McGregor et al. 1981). In female songbirds, social traits that lead to success in competition for territories and mates are less well studied than those of males. They represent a promising area for future study.

Estimates of effective population size based on the social mating system may differ from those based on the genetic mating system (Parker and Waite 1997). Such a bias is most likely when a few mated males monopolize most extrapair matings or when unmated males sire many young. Neither of these situations applied to song sparrows on Mandarte, although unmated males on Mandarte Island did sire a few young. Extrapair fertilization was common, but it was mostly by mated neighbors. Therefore, estimated effective population sizes for the social and genetic mating systems were virtually identical. Parker and Waite (1997) report similar results for purple martins and blue tits. We expect effective population sizes of other socially monogamous birds to be relatively unbiased by mate infidelity.

Social influences on numbers have been established clearly in some species of birds (e.g., great tits: Krebs 1971; Drent 1984; willow ptarmigan: Hannon 1983; Watson 1985; Capricorn silvereyes: Kikkawa 1980; Catterall et al. 1982). Individual social traits also have strong effects on fitness in the song sparrow. Effects of dominance on juvenile survival in the 1980s were strong. We suspect that they were even stronger from 1992 through 2001 when survival of juvenile females was low (see chapter 4), but we did not specifically study social dominance during this period.

While social competition is almost universal among vertebrates (e.g., Wilson 1975), its intensity might be reduced in small and isolated populations. However, many endangered bird populations live on tiny islands where high-density conditions still prevail (Bell and Merton 2002) and where social mechanisms can still influence numbers.

Notes

1. Annual reproductive success of females that did/did not switch mates was estimated for 25 years with mate switching. We compared mean numbers of young reared to fledging and 24 days of age in all clutches laid by switching/nonswitching females in the same year, that is, correcting for effects of population density on reproductive success. We excluded females that bred only once and females involved in feeding and male removal experiments in 1979 and 1985. Females that switched mates reared on average 0.58 more fledglings (SE = 0.27) and 0.30 more independent young (SE = 0.26) than did females that bred only with one male.

2. Pearson correlations: $r = 0.45$, $p = 0.02$, $n = 27$ years, for numbers of male floaters versus male density; $r = 0.32$, $p < 0.10$, $n = 27$ years, for the proportion of floaters versus male density.

3. Data on polygyny were analyzed by logistic regression. For the sexes analyzed together, sex ratio ($p < 0.001$), number of breeding females ($p < 0.001$), age (categorical classes from 1 to 5+ years, all $p \leq 0.02$) and one of four age $\times$ sex interaction terms (age 2 contrast, $p < 0.001$) all contributed to the probability of polygyny (experimental animals excluded, each pair treated as an independent observation, logistic regression $G = 98.4$, df = 11, $p < 0.001$, McFadden's $rho^2 = 0.13$). When we included experimental birds and coded mate removal or feeding as treatment effects, these results were strengthened. In this latter model, two of four age $\times$ sex interaction terms (age 2 contrast, $p < 0.001$; age 3 contrast, $p < 0.02$) contributed to our ability to predict the probability of polygyny.

For females ($n = 813$ pairings of known-age females, 66 of them polygynous), population sex ratio ($p < 0.001$), number of breeding females ($p = 0.003$), and age (all $p < 0.02$) were all related to the probability of polygyny (model log-likelihood = 50.28, df = 6, $p < 0.001$, McFadden's $rho^2 = 0.11$). Experimental birds were excluded from this analysis. When "experiment" was included as a categorical dummy variable, the results were strengthened for all variables.

Similarly, for males ($n = 874$, pairings of known-age males, 30 of them polygynous), the pattern was roughly opposite that for females; 2-year-olds were more likely, and other age-classes less likely, to be polygynous. In this case, however, only population sex ratio ($p < 0.001$) and number of breeding females ($p = 0.003$) were significant predictors of polygyny (model log-likelihood = 19.17, df = 2, $p < 0.001$, McFadden's $rho^2 = 0.07$). However, this model was improved significantly by adding age class ($\Delta AIC = 7.26$, df = 1, $p < 0.01$, McFadden's $rho^2 = 0.12$). With "experiment" was included as a categorical dummy variable, the results were again strengthened.

4. Numbers of young reared to fledging and 24 days of age were compared to the mean numbers of young reared by monogamous females in the same breeding year. Data for 1979, when polygyny was created experimentally, were excluded. In 1985, when feeding had a significant effect on reproductive success, each polygynous female was compared to the mean for her treatment group. In 1988, when feeding did not affect reproductive success, the overall mean was used. Average deviations were all negative, and more strongly negative for females that shared a social mate for the entire year than for females that were paired for part of the year.

7 The Genetic Consequences of Small Population Size: Inbreeding and Loss of Genetic Variation

Lukas F. Keller, Amy B. Marr, and Jane M. Reid

After the severe winter storm of 1989, the population of song sparrows on Mandarte Island was reduced to 11 individuals. Sparrows born in 1989 therefore had few mates to

choose from, and inbreeding among close relatives increased. For example, in the spring of 1990, female 54533 paired with her brother 69520. Both birds hatched in 1989, but from different clutches. Would song sparrows like these show the same detrimental effects of inbreeding seen in domestic animals? Alternatively, would fitness variation caused by competition, predation, and cowbirds overshadow any effects of inbreeding?

7.1. Genetic Consequences of Small Population Size: A Brief Historical Perspective

When agriculturalists breed domestic animals and plants to favor desired traits, they almost always work with small populations. Therefore, the genetic consequences of small population size have been of interest for a long time. Inbreeding depression, the reduced performance of individuals with related parents, has been studied for more than 200 years. Inbreeding occurs most frequently within small populations, since options to mate with nonrelatives are limited. Pioneering experiments on inbreeding by Kölreuter (1766), Knight (1799), and others were followed by Charles Darwin's book on the consequences of inbreeding and crossbreeding in plants (Darwin 1876). Darwin reported experiments with no less than 57 plant species! However, Darwin was unaware of Mendel's recent discovery, in 1866, that the amount of heterozygosity halves in each generation of self-fertilization. Darwin's interpretations of the causes of inbreeding depression therefore remained vague.

The early part of the twentieth century saw a series of in-depth experiments on inbreeding with both animals and plants (for details, see chapter 2 in Wright 1977). Thanks to the rediscovery of Mendel's laws of inheritance in 1900, these experiments eventually led to a theory of inbreeding. Two principal hypotheses to explain inbreeding depression emerged: dominance and overdominance (Crow 1952). The dominance hypothesis proposes that inbreeding increases the occurrence of deleterious recessive alleles in a homozygous state. The overdominance hypothesis suggests that inbreeding depression results from a reduced occurrence of superior heterozygote genotypes. These two hypotheses remain our primary explanations for inbreeding depression today. Overall, the dominance hypothesis seems to best explain observed patterns of inbreeding depression, but overdominance is involved in some cases (Charlesworth and Charlesworth 1999).

While debate on the genetic basis of inbreeding depression contin-ues, inbreeding in animals nearly always reduces fitness (Wright 1977). Early in the development of conservation biology, this evidence from animal husbandry, and parallel evidence from captive propagation of wild animals (Ralls et al. 1979), led to concerns about inbreeding de-pression in the wild (e.g., Soulé 1980). Since inbreeding and loss of ge-netic variation go hand in hand, conservation biologists highlighted how these processes might harm the small populations typically seen in en-dangered species (Frankel 1974; Ralls et al. 1979). This concern later became a central element in the "small population paradigm," a body of theory that links small population size to extinction probability (Caughley 1994; see chapter 1).

However, while evidence from zoo populations supported the view that inbreeding effects were a conservation concern (e.g., Ralls et al. 1988), comparable data from natural populations were scarce. Several early studies of inbreeding effects in the wild failed to detect inbreed-ing depression (see chapters in Thornhill 1993). Furthermore, no nat-ural population had then been observed to decline or go extinct due to inbreeding (Caro and Laurenson 1994). Several conservation biologists therefore began to question whether inbreeding was a major force in the wild (e.g., Harcourt 1991; Shields 1993). Some even considered that inbreeding depression was only of minor importance for wild pop-ulations (e.g., Caro and Laurenson 1994; Caughley and Gunn 1996). This view, however, is hard to reconcile with the extensive evidence from agricultural and laboratory systems that inbreeding is a serious problem. To resolve this dilemma, it was suggested that inbreeding avoidance and genome purging might reduce inbreeding and inbreed-ing depression in the wild (e.g., Ralls et al. 1986; Hedrick 1994).

Inbreeding avoidance is widespread in the animal kingdom (Pusey and Wolf 1996). If relatives rarely meet as adults, or if they actively avoid mating when they do meet, close inbreeding will be rare in the wild (Ralls et al. 1986). Inbreeding among distant relatives (e.g., sec-ond or third cousins) might occur commonly, but fitness costs of in-breeding at this level would be hard to detect even in exhaustive field studies.

The magnitude of inbreeding depression in the wild might be low because genome purging reduces the genetic load. The idea here is that the process that causes inbreeding depression, the homozygous ex-pression of deleterious recessive alleles, will purge these same deleteri-ous recessive alleles from the gene pool. When population bottlenecks cause inbreeding, these deleterious alleles will be exposed to natural selection and, ultimately, will be lost (Lynch and Walsh 1998). Since many natural populations experience bottlenecks during their evolutionary history, genome purging could reduce the magnitude of inbreeding depression in the wild (Hedrick 1994; Hedrick and Kalinowski 2000).

In the early 1990s, when this debate was in full swing, detailed analyses of inbreeding depression in the Mandarte Island song sparrow population became possible. Almost all song sparrows living on Mandarte had been individually color-ringed over nearly 20 years. For two birds, we knew 17 generations of their ancestors in 1993. This deep pedigree allowed us to calculate inbreeding coefficients accurately (see also section 7.8). These calculations were further aided by the low immigration rate (see chapter 4), which kept the pedigrees of most individuals complete. Further, our life-history data covered not only the life span of each individual bird, but also that of its offspring and grand-offspring. Hence, our study provided an ideal testing ground for the effects of inbreeding in wild populations. In this chapter, we document the incidence of inbreeding on Mandarte in relation to population bottlenecks, describe its effects on fitness, and place our results in the context of other recent studies of inbreeding in the wild. We close with a summary of what we do and do not know about inbreeding in natural populations.

7.2. The Occurrence of Inbreeding on Mandarte Island

Given the isolation and small size of the Mandarte population (see chapters 2 and 4; see also figure 4.1) and the dispersal patterns of the sparrows (Arcese 1989a; see chapter 4), inbreeding is likely to occur regularly. Indeed, on average, more than half of all the matings on Mandarte were between known relatives (Keller 1998; Keller and Arcese 1998).

However, the occurrence of inbreeding varied substantially throughout our study. The 1989 crash changed the frequency of inbreeding on Mandarte dramatically (figure 7.1). First, the severe winter storm that caused the crash selected strongly against inbred birds (Keller et al. 1994; see below), and consequently, the average inbreeding level in the population dropped between 1988 and 1989. Thereafter, inbreeding levels rose dramatically, and the average mates were more closely related than first cousins within 2 years. Since 1993, the average level of inbreeding has remained at or slightly above the first-cousin level ($f = 0.0625$), except for a dip in 2000.

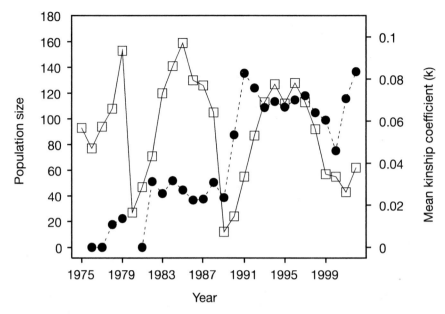

Figure 7.1. Year-to-year variation in inbreeding in song sparrows expressed as the mean kinship coefficient (k) of all pairs breeding in a given year (solid circles). The size of the breeding population in each year is represented by open squares. We calculated Wright's (1969) coefficient of inbreeding, f, relative to the baseline population in 1975 using PROC INBREED in SAS. Similarly, we calculated the coefficient of kinship, k, between the male and a female of a pair, which equals the inbreeding coefficient of their offspring (Crow and Kimura 1970). Details of the pedigree construction and the calculation of inbreeding and kinship coefficients are given in Keller (1998) and Marr et al. (2002). To standardize for differences in pedigree depth across the study period, we included only data from birds for which all four grandparents were known in all the analyses in this chapter.

When we first noted a strong increase in average inbreeding following the crash, we were not surprised. An increase in inbreeding following a bottleneck is expected from theory (Wright 1977). However, our colleague Mark Beaumont (currently at the University of Reading, UK) pointed out that we should have been surprised. Increased inbreeding following a bottleneck is only expected when a population is closed, that is, does not receive immigrants. Mandarte, however, receives a steady trickle of immigrants (see figure 4.6; see also chapter 8), and this changes the theoretical expectation completely. Under the assumption that a constant number (rather than a constant proportion) of immigrants breeds on Mandarte each generation, the proper expectation is that the average level of inbreeding should not change despite the dramatic bottleneck (M. Beaumont, personal communication). Mark's seemingly counterintuitive prediction stems from the fact that, with a constant number of immigrants, immigration and genetic drift should cancel each other out exactly: When population size is low, genetic drift is most pronounced, but so is the proportional effect of the immigrants. The opposite results apply when population size is large. Consequently, one should not expect a change in average inbreeding. Although the number of immigrants to Mandarte was not precisely constant, it varied relatively little and was unrelated to the population size on Mandarte (Marr et al. 2002; see also chapter 8). If anything, immigration was higher immediately after the 1989 crash (see figure 4.6; Keller et al. 2001), suggesting that the bottleneck did not cause the observed increase in inbreeding.

We therefore investigated whether other mechanisms could explain the increased inbreeding after the bottleneck. The most likely explanation is that the crash survivors in 1989 were not a random sample of the population, but were more closely related than expected by chance. This increased relatedness would translate to an increase in inbreeding in the following generations. Indeed, the average degree of relatedness among bottleneck survivors was 0.038, while the average of all breeding birds in 1988 was 0.027 (Keller et al. 2001).[1] This increased relatedness among bottleneck survivors occurred despite the fact that the 1989 crash selected against inbred birds. Such increased relatedness of survivors might be a common outcome of bottlenecks in nature if traits that promote survival under extreme environmental conditions are heritable. For example, wing length was correlated positively with female survival over the 1989 crash (Rogers et al. 1991) and is heritable on Mandarte (Schluter and Smith 1986). Selection on a trait such as wing length could lead to the survival of relatives and increased inbreeding subsequently.

Our data also suggest that song sparrows did not avoid inbreeding (Keller and Arcese 1998), a somewhat surprising result that we return to further below. Thus, finite population size rather than patterns of mate choice seems to be the key factor affecting annual levels of inbreeding on Mandarte Island.

7.3. Inbreeding Depression

Since inbreeding occurred commonly on Mandarte, we were able to investigate whether it was associated with a decline in fitness. We show below that inbreeding depression was prevalent among song sparrows on Mandarte and that its effects were as strong as seen in laboratory and agricultural populations. We first examined inbreeding depression among Mandarte song sparrows with data up to 1995 (Keller 1998). Here, we revisit the effects of inbreeding on several traits with the considerably larger data set available up to 2002. We also touch on some recent work on the effects of inbreeding on immune function and the expression of a sexually selected trait, male song repertoire size.

Inbreeding Depression in Survival

Inbreeding depression during the severe environmental stress that accompanied the 1989 population crash (Keller et al. 1994) was so strong that no bird with an inbreeding coefficient greater than 0.048 survived the crash (figure 7.2). Inbreeding depression in survival was also evident across all years of the study combined. Inbred song sparrows survived less well from independence onward.[2] An increase in inbreeding of 10%, that is, a change in inbreeding from $f = 0$ to $f = 0.1$, resulted

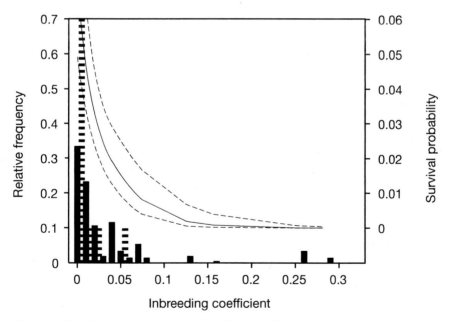

Figure 7.2. The distribution of inbreeding coefficients among song sparrows that died during the 1989 crash (solid bars) and those that survived (striped bars). The shape of the relationship between survival probability and inbreeding is shown by the cubic spline fit (solid line) with 95% confidence interval (dashed lines).

on average in a 7% decrease in annual survival. It is often thought that inbreeding depression is particularly strong early in life and less pronounced late in the life cycle (e.g., Keller 1998). Therefore, we also analyzed juvenile survival (from independence of parental care to 1 year of age) separately from adult survival (from age 1 onward).

These results are interesting for two reasons. First, the degree of inbreeding depression in juvenile survival (regression slope = 1.24) is slightly lower than among adults (slope = 1.42), although the latter is not significantly different from zero due to the much smaller sample size.[2] Thus, inbreeding depression in survival of song sparrows was not more pronounced in early life stages.

Second, adding an effect of sex to the model relating inbreeding to adult survival revealed that there was statistically significant inbreeding depression only among males.[2] Adult females apparently did not suffer any inbreeding depression in survival. This result surprised us because analyses up to 1995 (Keller 1998) did not reveal this difference between males and females. We noted in chapter 4 that some attributes of the population changed after 1989; in particular, the survival of juvenile females became poor. We therefore analyzed inbreeding depression in adult survival separately for these two periods.[2] Indeed, the difference in inbreeding depression between the sexes is only evident after 1989. In fact, the difference in survival between adult males and females irrespective of inbreeding (see figure 3.5) is apparent only after 1989. While the cause of the reduced female survival after 1989 is unknown, inbreeding does not seem to have been the main cause. We could not analyze sex-specific inbreeding depression among juveniles because we could not tell males and females apart until they were fully grown (see chapter 2).

Inbreeding Depression in Annual Reproductive Success

Inbreeding also affected reproductive success on Mandarte Island.[3] Several components of annual reproductive success (ARS) were depressed in inbred birds, and as was the case for survival, males and females differed in their patterns of inbreeding depression (table 7.1). Both inbred males and inbred females raised fewer young to independence. A 10% increase in inbreeding (i.e., $f = 0$ to $f = 0.1$) resulted in an 20% and 17% decline in the number of independent young produced by males and females, respectively. In females, inbreeding depression in ARS was primarily a consequence of the reduced hatching rates among inbred mothers. Inbreeding did not affect the later stages of the life cycle, but the pronounced

Table 7.1. Inbreeding depression in ARS in female and male song sparrows.

Females			
Date of first egg	Inbred females lay later	28.3 (9.04, 47.5)	$P = 0.004$ $n = 280$
Number of breeding attempts	No effect	-0.48 (-1.13, 0.17)	$P = 0.15$ $n = 280$
Number of eggs laid	No effect	-0.26 (-0.99, 0.46)	$P = 0.48$ $n = 280$
Number of young raised to independence	Inbred females raise fewer young	-1.83 (-3.29, -0.37)	$P = 0.014$ $n = 280$
Percent eggs that hatch	Inbred females experience a low hatching rate	-2.57 (-4.88, -0.27)	$P = 0.03$ $n = 275$
Percent hatchlings that fledge	No effect	-1.39 (-3.77, 0.98)	$P = 0.25$ $n = 265$
Males			
Date of first egg	Females mated to inbred males lay later	25.7 (1.34, 50.1)	$P = 0.04$ $n = 389$
Number of breeding attempts	Inbred males have fewer attempts	-2.01 (-3.26, -0.75)	$P = 0.002$ $n = 389$
Number of eggs laid	Females mated to inbred males lay fewer eggs	-1.79 (-3.23, -0.35)	$P = 0.015$ $n = 389$
Number of young raised to independence	Inbred males raise fewer young	-2.23 (-3.88, -0.58)	$P = 0.008$ $n = 389$
Percent eggs that hatch	No effect	1.53 (-0.87, 3.92)	$P = 0.21$ $n = 283$
Percent hatchlings that fledge	No effect	-1.14 (-3.48, 1.19)	$P = 0.72$ $n = 274$

Note. Sample sizes are the numbers of individual birds in each analysis.

inbreeding depression in hatching rates translated into a lower number of independent young raised. In males, inbreeding depression also occurred primarily during the early stages of the reproductive cycle. Pairs with an inbred male laid fewer eggs per season. This was because they started breeding later and therefore made fewer nesting attempts per season. Therefore, although a different stage of reproduction was affected than in females, inbred males also raised fewer independent young.

Inbreeding Depression in Lifetime Reproductive Success

Lifetime reproductive success (LRS, the number of breeding offspring that an individual produced over its lifetime) integrates reproductive success and survival (see chapter 9). LRS showed significant inbreeding depression, and mirroring the case for survival, inbreeding depression affected males more than females.[4] Males exhibited significant inbreeding depression: A 10% increase in inbreeding decreased LRS on average by 25%. Female LRS, on the other hand, was unaffected by inbreeding, despite the fact that female ARS showed significant inbreeding depression. However, the confidence intervals around the estimates of inbreeding depression in female LRS are wide and include the estimate for males.[4] Thus, while we do not have statistical support for inbreeding depression in female LRS, the data cannot be taken as evidence that inbreeding depression was truly absent among females.

Overall, inbred males showed inbreeding depression in survival and in ARS, which together led to reduced LRS. Inbred females, on the other hand, suffered significant inbreeding depression in reproduction, but they survived as well as their outbred counterparts (except up to 1989). As a result, they did not exhibit significant overall inbreeding depression in LRS. We discuss other factors contributing to LRS in more detail in chapter 9.

Our results here differ somewhat from those in Keller (1998). This is because Keller used a different definition of LRS (the total number of independent young instead of the total number of breeding offspring used here), alternative statistical techniques, and only data up to 1995. The first and third differences account for the fact that female LRS no longer shows inbreeding depression in the present analyses, whereas it did in Keller (1998). In males, however, inbreeding depression in ARS was weaker in the earlier data set than we observed here, although statistical support for this pattern is not strong ($p = 0.05$). Thus, inbreeding depression in male LRS has only become evident with the extended data set. This leads to a sobering insight: It took more than the 21 years of data analyzed in Keller (1998) to detect substantial inbreeding depression in male LRS on Mandarte! This reinforces the observation that detecting inbreeding depression in natural populations is difficult (Kalinowski and Hedrick 1999) and requires large sample sizes (Kruuk et al. 2002). While no less sobering, this insight is not new: "The evil results from close inbreeding are difficult to detect, for they accumulate slowly" (p. 92 in Darwin 1868).

Physiological Explanations of Inbreeding Depression

We have seen that inbred birds survive less well (males) and reproduce less well (males and females). What are the physiological mechanisms that cause these patterns? Modern studies of variation in life history traits look for physiological and developmental mechanisms that might underlie both reproduction and survival (Ricklefs and Wikelski 2002; Nowicki et al. 1998).

Many physiological mechanisms could contribute to the inbreeding depression we observed on Mandarte Island. One is the immune system. We had two reasons to be particularly interested in whether variation in immune function could cause inbreeding depression. First, since epidemics can alter population dynamics, a decline in immunity with inbreeding could affect the dynamics and persistence of the small and isolated populations in which inbreeding is likely (O'Brien and Evermann 1988). Second, disease is one possible cause of the population decline on Mandarte in 1999 (see chapter 4). If inbred individuals have less active immune systems due to reduced heterozygosity at loci promoting immunity, they could be less able to withstand infectious diseases and debilitating parasites. As a result, they might survive and reproduce less well.

We recently tested these ideas experimentally by measuring the immune responses of inbred and outbred song sparrows on Mandarte Island. We injected one patagium (the flap of skin in front of the upper arm bone of all flying birds) of both adult and nestling song sparrows on Mandarte with a small amount of phytohemagglutinin, a plant mitogen (i.e., a substance that causes cells to divide; Reid et al. 2003a). Song sparrows are unlikely to ever have been exposed to this mitogen, and they mount an immune response, which is evident as a local swelling at the site of the injection. The thicker this swelling is, the stronger the immune response. We found that inbred individuals showed a much weaker immune response than did outbred individuals, and we saw the same effect when we repeated this experiment three separate times. Interestingly, we also found that immune response in chicks varied with their mother's inbreeding level; chicks of inbred mothers showed a weaker immune response than did chicks of outbred mothers (Reid et al. 2003a).

In a further study, we found that inbred males had smaller song repertoires than did outbred males (Reid et al., 2005b), and males with smaller song repertoires were less likely to obtain a mate in their first breeding year (see figure 6.5; Reid et al. 2004). In addition, males with larger song repertoires had a higher LRS (see figure 6.5; Reid et al. 2005a). Furthermore, females mated to first-year males with smaller song repertoires started breeding later in the season (see figure 6.5; Reid et al. 2004), a pattern reflected in the finding that females mated to more inbred males bred later in the season (table 7.1). Taken together, these results go a long way toward explaining why inbreeding reduces male LRS.

7.4. Variation in Inbreeding Depression over Time

The magnitude of inbreeding depression in several traits (onset of laying, hatching success, juvenile survival) varied strongly over time (figure 7.3). The time frame over which this variation occurred was far too short for it to be explained by genetic processes such as mutations, purging, or immigration. In other studies, changing environmental conditions have been identified as the source of short-term variation in inbreeding depression (e.g., Keller et al. 2002).

Environmental conditions also account for some of the variation in the magnitude of inbreeding depression on Mandarte. For example, inbreeding depression in male ARS (number of independent young raised) depends on spring temperatures. When the months January through April are warm, inbred males produce as many independent young during the entire breeding season as do outbred males (figure 7.4). However, after cold springs, inbred males ($f = 0.25$) produce on average 46% fewer independent young. Thus, inbreeding depression is evident only when environmental conditions are poor. Another trait where the magnitude of inbreeding depression and environmental conditions are linked is hatching rate. The effect of female inbreeding on hatching rate described above (table 7.1) depends on both rainfall during the laying season and cowbird parasitism (A.B. Marr, unpublished observations). However, for most traits, we have not been able to identify an environmental variable that might drive the temporal variation in inbreeding depression.

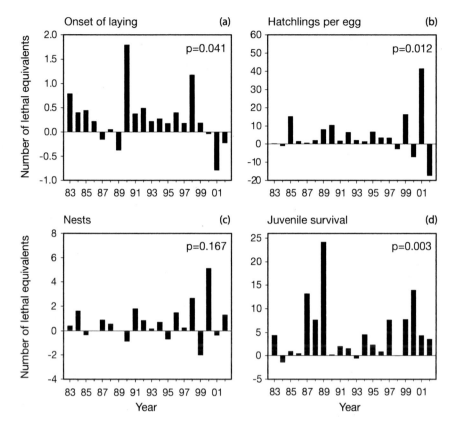

Figure 7.3. Variation in inbreeding depression, expressed as number of lethal equivalents, over time. Lethal equivalents are the number of deleterious genes whose cumulative effects are the equivalent of one lethal. They provide a standardized measure of the magnitude of inbreeding depression (e.g., Keller and Waller 2002). Here we estimated the number of lethal equivalents per gamete for a range of life-history traits. In (c), "nests" refers to the number of nesting attempts per breeding season.

7.5. Did Genome Purging Reduce the Magnitude of Inbreeding Depression on Mandarte?

Given the repeated bottlenecks that have occurred on Mandarte (see chapter 4), we might expect genome purging to have reduced the observed levels of inbreeding depression. This, however, is not the case (figure 7.5). The magnitude of inbreeding depression observed on Mandarte is almost identical to the average found in other studies (figure 7.5).

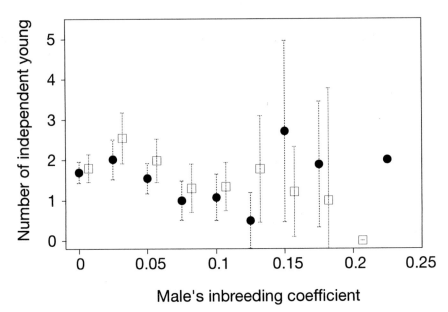

Figure 7.4. When the months January through April are warm (circles), inbred male song sparrows produce as many independent young during a breeding season as do outbred males. In cold springs, however, inbreeding depression in male annual reproductive success is pronounced (squares). Inbreeding coefficients of males were grouped into nine categories and spring warming into two categories for the purpose of this graph. The mean number of independent young produced per male is shown with 95% confidence intervals. Note: the statistical analysis corrected for effects not displayed in this figure. Environmental conditions affected the magnitude of inbreeding depression in male annual reproductive success (ARS). We used generalized linear mixed models (PROC GENMOD in SAS) to identify the environmental variables that affect inbreeding depression in male ARS (the number of independent young produced). Our approach is described in detail in Keller et al. (2002). The final model included the following significant main effects: density of males in a season, a male's age, spring warming, rainfall during the season (March through July), and a male's inbreeding coefficient. The only significant interaction term was between a male's inbreeding coefficient and warming ($b = -0.017$; 95% confidence interval, -0.03 to -0.004; $p = 0.01$; $n = 389$) indicating that inbreeding depression was more severe when spring temperatures were cold. Spring warming is the cumulative degree days of warming in °C (Wilson and Arcese 2003). Higher values of spring warming indicate colder springs and vice versa.

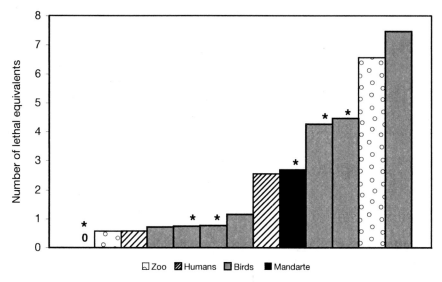

Figure 7.5. A comparison of the magnitudes of inbreeding depression among populations, expressed as the number of lethal equivalents. The estimate from Mandarte (black) is given together with all estimates available from wild bird populations (gray). The two most extreme values found in human populations (hatched) and zoo populations of mammals (dotted) are also given. One study estimated the number of lethal equivalents to be zero. Data from small or bottlenecked island populations are denoted with an asterisk. Based on data in Keller et al. (2002) and Kruuk et al. (2002).

Similarly, the magnitudes of inbreeding depression in other small or bottlenecked island populations of birds do not suggest that these populations were purged of substantial amounts of their genetic load (figure 7.5). Taken together, these results suggest (a) that the results obtained on Mandarte are not unusual, and (b) that genome purging has not reduced the genetic load to an appreciable degree from this natural population. Theoretical investigations and experimental evidence support the view that genome purging may not be as efficient in natural populations as is sometimes assumed (e.g., Hedrick 1994; Glémin 2003). In particular, recent work has shown that while genome purging can be very effective when inbreeding occurs through nonrandom mating (Crnokrak and Barrett 2002), it is unlikely to work well in small, random mating populations (Whitlock 2002; Glémin 2003). As described in section 7.2, the Mandarte song sparrows are exactly such a population (Keller and Arcese 1998).

In addition, immigration may explain the absence of substantial genome purging on Mandarte. A steady trickle of immigrants arrives on the island (see chapters 4, 8), and they bring with them new genetic variants. Is this gene flow sufficient to reintroduce genetic variation purged in bottlenecks? An analysis of genetic variation at eight microsatellite marker loci suggests that it is (Keller et al. 2001). We started to collect blood samples annually in 1987. Therefore, we were able to analyze how variation at these marker loci changed over the 1989 bottleneck.

As expected, several rare alleles were lost during the 1989 bottleneck. However, the steady trickle of immigrants re-established genetic variation in only 3 years (figure 7.6). This was true for two measures of genetic variation: the average number of alleles per locus and average expected heterozygosity. The latter measure combines information on both number of alleles and their relative frequencies and is therefore the better measure. This recovery of heterozygosity and lost alleles was because of immigration and not genetic drift or selection (Keller et al. 2001). We determined this by using pedigrees to identify every song sparrow that had an immigrant ancestor. Removing all birds with immigrant ancestors from the data set allowed us to quantify what

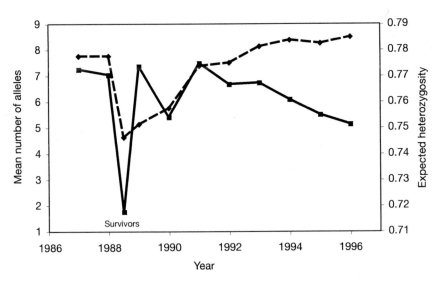

Figure 7.6. Average expected heterozygosity (squares and solid line) at six microsatellite loci and average number of alleles (diamonds and dashed line) at eight microsatellite loci from 1987 to 1996. Reprinted with permission from Keller et al. (2001).

would have happened on Mandarte in the absence of immigration: a decline of heterozygosity to much lower levels than with immigration (figure 7.6). Hence, immigration and not drift or selection accounted for the rapid increase in genetic variation on Mandarte after the 1989 crash (see also chapter 8).

It is very likely, therefore, that immigration explains in part the continued inbreeding depression in the serially bottlenecked Mandarte population. Note also that genetic variation was restored at the same time as average inbreeding increased dramatically (compare figures 7.1. and 7.6). This coincidence emphasizes that, with immigration, inbreeding and genetic variation are not as tightly linked as they are in closed populations, which form the basis of many theoretical models.

7.6. Why Do Song Sparrows Inbreed?

We have seen that song sparrows incur substantial costs when they inbreed, yet they do not seem to avoid mating with relatives despite these costs (Keller and Arcese 1998). This result seems inconsistent with evidence for inbreeding avoidance in other vertebrates (Pusey and Wolf 1996). We therefore searched carefully for evidence of inbreeding avoidance.

Song sparrows breeding for the first time generally pair with another first-year individual and tend to keep the same mate from year to year (see chapter 6). Thus, whether inbreeding is avoided or not is mainly decided in a bird's first winter. We therefore asked if siblings avoid each other's home ranges more than do nonsiblings, as they are becoming breeding adults. Avoiding a shared home range with a sibling could indicate inbreeding avoidance.

We used detailed observations of juvenile movement patterns recorded by Peter Arcese in 1982–1985. During those years, Peter recorded the exact location of every sighting of a juvenile bird on a map. From these maps, we calculated home ranges for every juvenile bird and asked how much the home ranges of siblings overlapped compared to those of nonsiblings.[5] Home ranges of siblings overlapped more than those of nonsiblings in the period following fledging (figure 7.7a). This is not surprising, since siblings originate from the same territory. However, over the winter and into spring, home range overlap decreased as birds dispersed, and the home ranges of siblings eventually overlapped the same amount as did ranges of unrelated juveniles. This result held true when we compared siblings from the same nest to sib-

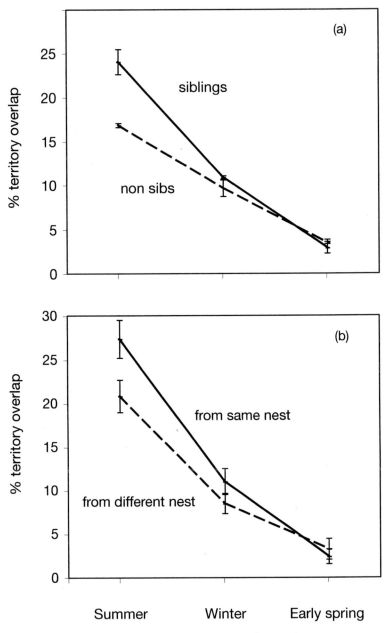

Figure 7.7. Percentage overlap in home ranges of juvenile song sparrows during three periods of their first year. On the x-axis, summer denotes the period from fledging to August 31; winter, from September 1 through January 15; early spring, January 16–April 30, when most birds had settled on territories to breed. Data are from 1982 through 1985 only: (a) siblings compared to nonsiblings, (b) siblings born in the same nest compared to siblings born in different nests.

lings hatched in different nests (figure 7.7b). Therefore, familiarity of nest mates did not affect home range overlap. In conclusion, home range overlap of siblings is indistinguishable from overlap of unrelated birds. This result suggests that song sparrows do not show dispersal behavior that would lead to inbreeding avoidance.

We asked more directly whether song sparrows avoided inbreeding (Keller and Arcese 1998). Here, we used three different approaches to compare observed mating patterns on Mandarte to those expected if birds mated at random. All three gave the same answer: Song sparrows on Mandarte Island mated with relatives as often as expected if matings occur randomly. It is always difficult to be certain that something does not occur, but the fact that three different approaches gave similar results suggested to us that inbreeding avoidance is simply absent (Keller and Arcese 1998).

Since inbreeding costs appear to be substantial in song sparrows on Mandarte, is there a theoretical reason to explain why they do not avoid mating with a relative? Classical theory in behavioral ecology (e.g., Krebs and Davies 1987) tells us that a trait should evolve only if the benefits outweigh the costs. We should therefore be able to predict whether sparrows should avoid inbreeding on Mandarte by comparing the costs and benefits of this behavior. We have a clear idea of the benefits of inbreeding avoidance, since these ought to be similar in magnitude to the degree to which inbreeding depresses fitness. The costs of inbreeding avoidance, however, are less clear. One possible cost of inbreeding avoidance is that a song sparrow does not mate in its first breeding year. Given that the sex ratio on Mandarte is often male biased (see chapter 3; see also figure 4.1) and that territory acquisition is competitive in both sexes (Arcese 1989a, 1989c), this assumption seems plausible.

If we assume that the costs of inbreeding avoidance equal the loss of one breeding season, a simple mathematical model suggests that the costs of inbreeding avoidance on Mandarte are larger than the benefits (Keller and Arcese 1998). Inbreeding avoidance is generally absent in more short-lived small passerines (great tits and collared flycatchers; for details, see Keller and Arcese 1998). The same model also helps to explain why long-lived birds do show inbreeding avoidance (e.g., acorn woodpecker, Koenig et al. 1998). Of course, this calculation is approximate, and we may have under- or overestimated the cost of inbreeding avoidance. For example, some unmated males do mate later in the season and/or acquire extrapair paternities (see chapter 6).

An alternative explanation for the absence of inbreeding avoidance is that song sparrows cannot recognize their relatives. This may sound heretical, given that some animals have fantastic abilities to recognize relatives (e.g., Ryan and Lacy 2003). A mechanism that songbirds might use to recognize relatives is song. In some songbirds, sons copy the songs of their fathers (e.g., North Island saddlebacks, Jenkins 1978; Darwin's finches, Grant and Grant 1996a), providing a ready signal of relatedness. However, male song sparrows learn most of their songs after they have dispersed from their natal territory, and fathers and sons do not share similar songs (e.g., Cassidy 1993; Nordby et al. 1999). Thus, one of the most obvious mechanisms for a songbird to recognize its relatives is not available to a song sparrow.

Although we do not know why song sparrows on Mandarte do not avoid inbreeding, our study emphasizes the point that inbreeding avoidance is not ubiquitous: Aristotle's story of the young stallion who, after being tricked into mating with his mother, hurled himself from a cliff (Zirkle 1952) does not reflect the levels of inbreeding avoidance in all of nature.

7.7. How Does Inbreeding on Mandarte Compare to Other Populations?

Generalizing from the results of a single study is always risky. One factor that might make generalizing the Mandarte results particularly problematic is that Mandarte is an island. Because it is also small, inbreeding might be more frequent on Mandarte than in most song sparrow populations. To what degree this is the case is not clear. First, Mandarte receives a steady trickle of immigrants, increasing the effective population size considerably over a truly isolated population. Second, inbreeding also occurs in mainland populations because song sparrows are generally philopatric. In her study of a mainland population of song sparrows in Ohio, Nice (1937) documented one brother–sister pair among 14 pairs with known genealogies. This corresponds to a rate of close inbreeding ($f = 0.25$) of 7%. On Mandarte, 3.8% of all matings are between close relatives (Keller and Arcese 1998). Thus, at least in one mainland population close inbreeding does not occur significantly less frequently than on Mandarte. This view is supported by the observation that average heterozygosity at microsatellite loci on Mandarte is not significantly lower than in mainland populations in British Columbia (L.F. Keller, unpublished data). It is likely, however, that more distant inbreeding would be more frequent on an island like Mandarte than in the average mainland population.

If inbreeding is indeed more frequent on Mandarte, it would increase the statistical power of analyses of inbreeding depression. It might also make inbreeding avoidance more costly because unrelated mates are harder to find, and it might lead to purging of some of the genetic load. While some of the genetic load may have been purged on Mandarte, comparisons with other studies suggest that it was not reduced substantially (figure 7.5).

7.8. Inbreeding Depression in Wild Populations: What We Know and What We Don't

Several studies have asked questions similar to ours in the last decade. Thus, we can now summarize what we know about inbreeding depression in the wild and, equally important, what we don't know.

It has become abundantly clear that inbreeding depression occurs commonly in nature and that it can be severe. Rules of thumb derived from agricultural systems (i.e., a 10% increase in f leads to a 10% decrease in fitness) also hold in natural populations, at least to within a factor of 2 or 3. Therefore, there is no reason to think that lab studies will give fundamentally different answers to field studies. It might, however, be that inbreeding depression is more pronounced in natural populations than in captivity. One study, for example, reported that inbreeding depression in mortality among mammals was on average nearly seven times higher in wild than in captive populations (data from 9 wild and 40 captive populations; Crnokrak and Roff 1999). Similarly, inbreeding depression in mating success of the African butterfly *Bicyclus anynana* was greatly accentuated in seminatural conditions (Joron and Brakefield 2003): Almost twice the genetic load was expressed under free-flying conditions than in a cage.

However, we feel that generalizations here are premature for two reasons. First, field studies might be subject to different biases than are lab studies. For example, field studies tend to be based on smaller sample sizes than are lab studies, and the traits under study are likely to show more environment-induced variation in the wild. Both factors would mean that statistical significance would be reached in the wild at greater magnitudes of inbreeding depression than in captivity. Since statistically significant results are more likely to be published, the apparently larger magnitude of inbreeding depression in the wild (Crnokrak and Roff 1999) may be a consequence of this bias.

Second, where inbreeding is calculated from pedigrees based on behavioral observations (as it was in our case), extrapair paternities will introduce errors in the pedigrees that will bias estimates of inbreeding depression downward and reduce the statistical power of the analyses (Marr et al. in press). On Mandarte, 28% of offspring are on average not sired by their social father (see chapter 6), and a simulation study suggests that this leads us to underestimate inbreeding depression by up to 50% (Marr et al. in press). If, as is likely, extrapair paternities are more common in the wild than in captivity, current comparisons might in fact underestimate the differences in inbreeding depression between natural and captive populations. Taken together, we see potential biases that would both decrease and increase the apparent differences between captive and wild studies. These biases need to be addressed in comparative analyses before meaningful quantitative generalizations can be made about the relative magnitudes of inbreeding depression in the wild and captivity.

It has been well documented that the magnitude of inbreeding depression varies between species and populations (Lynch and Walsh 1998). It is perhaps more surprising that the magnitude of inbreeding depression can vary greatly within populations from one year to the next. We still have a poor understanding of the processes that cause this variation. As ecologists, we expect poor environmental conditions to increase the degree of inbreeding depression, yet this is only sometimes true (Keller et al. 2002; Henry et al. 2003). Poor environmental conditions do not always lead to greater inbreeding depression (e.g., Cheptou et al. 2000; Dahlgaard and Hoffmann 2000). This may have to do with the mechanisms linking inbreeding depression and environmental condition. If, as is the case for male LRS among Mandarte song sparrows, inbreeding depression is mediated by differences in song repertoires, environmental conditions surely must have very different effects to a situation where inbreeding depression is mediated by immune function.

In addition, recent evidence from experiments in plants support the view that the variance observed in a trait sets an upper limit to the expression of inbreeding depression (Waller et al. in press). When the phenotypic variance of a trait is high, inbreeding depression can be greater than when phenotypic variation is low. Because changes in environmental conditions may increase or decrease the phenotypic variance, environmental stress can have opposing effects on the magnitude of inbreeding depression, explaining the equivocal results observed. Thus, phenotypic variance may be more important than "stress" per se in determining variation in inbreeding depression among environments.

When does inbreeding of the parent matter? Most studies look at the effects of parental relatedness, but does grandparental relatedness matter? On Mandarte, we observed strong inbreeding depression in hatching rates of inbred females (table 7.1), and the chicks of inbred mothers showed a reduced response to a novel immune challenge (Reid et al. 2003a). That is, grandparental relatedness matters a lot for these traits. Similar results have been found in the endangered New Zealand takahe (Jamieson et al. 2003), and the poultry literature reports several maternal inbreeding effects. Despite these recent advances, however, we are only beginning to understand how widespread maternal inbreeding effects are in nature, what makes a trait likely to show maternal or paternal inbreeding effects, and whether these parental effects persist throughout life.

Finally, we would like to know the conditions under which inbreeding affects population dynamics. We know that inbreeding can affect population dynamics in the wild (e.g., Glanville fritillary butterfly, Saccheri et al. 1998) and in the laboratory (e.g., Frankham 1998; Reed et al. 2003). However, we do not know how common this result is and what conditions make effects on population dynamics in the wild more or less likely. Until we better understand the importance of inbreeding effects for population dynamics, we might be well advised to heed the rule of thumb developed in agriculture. When creating inbred lines, up to 19 of 20 lines are likely to go extinct when they are severely inbred over several generations (Wright 1977).

Notes

1. The average kinship coefficient of bottleneck survivors ($k = 0.0375$, SD $= 0.0607$, $n = 11$) was significantly different from the average kinship of all breeding birds in 1988 ($k = 0.0267$; SD $= 0.0464$; $n = 99$; sign test, $p = 0.009$). See Keller et al. (2001) for details. Note that the average kinship here is the average between all members of the population irrespective of gender, not the average between mating partners as shown in figure 7.1.

2. Survival analyses: Our estimates of survival are estimates of local survival, since birds that disappeared from Mandarte Island could have emigrated or died. While some juveniles are known to have dispersed (see chapter 4), dispersal is unlikely to have caused substantial biases in our analyses (Keller 1998). To quantify inbreeding depression in survival, we used discrete-time proportional hazards models following Keller (1998). The models were stratified by year of birth and age. This allowed us to quantify the decrease in annual survival with increased inbreeding among same-aged birds alive at the same time, thus minimizing the possibility that other variables (e.g., year or age effects) might contribute to the observed patterns. Because independence from parental care was monitored only from 1981 onward, analyses of juve-

nile survival included only years after 1981. Analyses of adult survival covered the entire study period (1975–2002). Birds still alive in 2002 were entered in the analyses as right-censored observations of survival time. Sample sizes differ between analyses because uninformative strata (where either all birds survive or all die) have to be excluded from the analyses (Heisey 1992). Positive parameter estimates indicate positive effects of this variable on the probability of death and thus negative effects on survival.

a. Survival from independence to death: Since date of birth is known to affect juvenile survival, we included that variable as a covariate in these analyses. Inbreeding significantly decreased survival ($b = 1.34$; 95% confidence interval, 0.40–2.29; $p = 0.006$; $n = 1735$ birds).

b. Survival from independence to 1 year of age, with date of birth as a covariate: Inbreeding significantly decreased juvenile survival ($b = 1.24$; 95% confidence interval, 0.07–2.41; $p = 0.042$; $n = 1735$ birds).

c. Survival from 1 year of age to death: There was no significant effect of inbreeding ($p = 0.09$). However, the magnitude of inbreeding depression in survival ($b = 1.42$; 95% confidence interval, -0.16–3.01; $n = 632$ birds) was greater than the one observed for juveniles. The large confidence interval suggests that this analysis lacks statistical power. While it does not indicate inbreeding depression, it does not prove its absence (Hoenig and Heisey 2001).

d. Survival from 1 year of age to death included an effect for sex differences: Inbreeding depression in survival was evident in males but not females, as indicated by a significant interaction between sex and inbreeding coefficient ($p = 0.035$, $n = 613$ birds). While inbreeding depression was not significant among adult females ($b = -1.12$, $p = 0.45$, $n = 250$), it was strong in adult males ($b = 3.44$; 95% confidence interval, 1.10–5.57; $p = 0.003$; $n = 363$). The differences in inbreeding depression between males and females, and the main effect of sex, were only statistically significant from 1989 onward (sex $\times$ inbreeding coefficient: $p = 0.05$; sex: $p = 0.02$) but not in 1975–1989 (sex $\times$ inbreeding coefficient: $p = 0.20$; sex: $p = 0.40$).

3. Inbreeding depression in ARS was analyzed using generalized linear mixed models (GLIMMIX macro and PROC GENMOD in SAS Version 8). Appropriate error distributions and link functions were chosen as follows: date of first egg (normal error, identity link), number of breeding attempts (Poisson error, log link), number of eggs (negative binomial error, log link), and percentages hatched and fledged (binomial error, logit link). We used female identity as a random factor to account for the multiple measure of ARS from most birds from different breeding seasons. Data points from birds subjected to experiments were excluded.

4. Inbreeding depression in LRS was analyzed using generalized linear models (PROC GENMOD in SAS). We found that a negative binomial dis-

tribution fitted the LRS data best, in line with empirical work on family size in humans and theoretical expectations (Cavalli-Sforza and Bodmer 1971). Individuals who were 5 years or younger in 2002 were excluded from the analyses because their LRS may still increase substantially. For reasons discussed in chapter 9, excluding birds that were involved in experimental manipulations can make the interpretation of LRS analyses more difficult. Rather than excluding such birds from the data set, we included the type of experiment as a categorical variable in the statistical models. Because LRS varies greatly among cohorts (see chapter 9), we also included year of birth as a categorical variable in all the models.

Males showed significant inbreeding depression in LRS ($b = -2.90$; 95% confidence interval, -5.63 to -0.18; $p = 0.04$; $n = 348$), while female LRS was apparently unaffected by inbreeding ($b = -0.85$; 95% confidence interval, -3.26 to 1.56; $p = 0.49$; $n = 268$). Note that the confidence interval for females includes the estimate for males.

5. Home range overlap: We divided each year's data into three periods, summer (June through August), winter (September through January), and spring (January through April). Detailed observations of the location of juveniles on the island between 1982 and 1985 were recorded on a map and later digitized. We calculated home range sizes (in hectares) for each individual in the data set using minimum convex polygons. For each pair of individuals in the data set, we then calculated the overlap of their respective home ranges (in hectares). The percentage overlap was calculated as twice the area of overlap divided by the sum of the two home ranges. Estimates of home range size and overlap were obtained using a program written by John Cary (Dept. of Wildlife Ecology, University of Wisconsin-Madison).

8 Immigrants and Gene Flow in Small Populations

Amy B. Marr

*Among the 12 song sparrows living on Mandarte Island af-
ter the severe storm of February 1989 was wo.gm, a female
born on nearby Rum Island who had immigrated the previous fall. She lived for only one breeding season and reared
four offspring during her short life span. All her offspring survived to the next year, and three had long and successful
lives. From 1990 to 1996, these three birds reared a remarkable total of 73 independent offspring. However, like the
descendants of many immigrant song sparrows to Mandarte
Island, the grand-offspring of wo.gm were less remarkable
than her offspring. Only 28 of her 73 grand-offspring survived to age 1, and these individuals had lower breeding*

> *success than did their contemporaries. Yet the sheer number*
> *of grand-offspring descended from wo.gm ensured that her*
> *genes were passed to a third generation and beyond.*

In chapter 7, we showed that individuals in small, isolated populations are more likely to inbreed and experience inbreeding depression. We also noted that isolated populations tend to lose the genetic variation that is vital for adapting to new environmental challenges and diseases. Studies of natural populations suggest that gene flow from immigrants can reverse the genetic deterioration of inbred populations; this is called *genetic rescue* (Vilà et al. 2003; Tallmon et al. 2004).

Management agencies sometimes introduce individuals from other populations to try to alleviate inbreeding depression and boost fitness. In one example, male adders were translocated to an isolated population in Sweden that had small litter sizes, many deformed and stillborn offspring, and low genetic heterozygosity (Madsen et al. 1999). The new individuals sharply increased the viability of offspring and genetic variation and reversed the population decline.

Despite their potential benefits, translocation programs are controversial. Translocated individuals can spread disease, increase competition, disrupt social groups (e.g., Cunningham 1996; Miller et al. 1999), and produce offspring that are either too fit due to heterosis or unfit due to outbreeding depression. In this chapter, I discuss the causes of heterosis and outbreeding depression and show how these genetic phenomena can affect efforts to rescue small populations.

8.1. Heterosis and the Spread of Immigrant Genes

Crosses between populations of a species sometimes yield first-generation progeny (F_1 generation) that are more fit than are purebred offspring from within either population. The botanist George H. Shull (1914) introduced the term *heterosis* to describe this hybrid vigor. Heterosis occurs in the offspring of crosses between populations because recessive deleterious alleles contributed by one parent are masked by normal alleles from the other parent (dominance). In addition, the offspring of crosses between genetically divergent populations may be more heterozygous at loci where heterozygotes have superior fitness (overdominance; Crow 1948).

It is often said that heterosis is the reverse of inbreeding depression (e.g., Falconer and Mackay 1996), but this explanation of heterosis has created confusion in the literature. The frequencies of deleterious alleles are likely to differ among populations due to selection and random drift (Crow 1948; Whitlock et al. 2000). Therefore, matings between populations can yield offspring that show heterosis, even if offspring from the most inbred and outbred matings within populations perform equally well.

Recent theoretical and empirical studies have demonstrated that heterosis can be crucial to the management of small populations because it affects the competitiveness of immigrant genes (e.g., Ingvarsson and Whitlock 2000; Ebert et al. 2002). In one such study, Saccheri and Brakefield (2002) established six laboratory populations of the squinting bush brown butterfly. Each was composed of 29 purebred families from matings between resident males and females and one hybrid family from a mating between a resident male and an immigrant female. Four generations later, the contribution of immigrant founders to the six gene pools was 2–71 times greater than the average contribution of resident founders. Heterosis could accelerate the genetic rescue of a population by improving population productivity (Richards 2000), but high levels of gene flow are not necessarily desirable (see below).

8.2. Gene Swamping

A high level of gene flow from immigrants to a population can lead to lost genetic variance at loci under selection. This is called *gene swamping* (Lenormand 2002). The genetic restoration program for the endangered Florida panther, a subspecies of the cougar, has been controversial because of concerns about gene swamping. To reverse symptoms of inbreeding depression in the tiny Florida panther population ($N = 30$–50 adults), eight female cougars were moved from Texas to Florida (Maehr and Lacy 2002). In less than 5 years, the Texas genes reached the target level of 20%, and successful genetic restoration was proclaimed (McBride 2000; Jansen and Logan 2002). By 2001, however, Texas genes had exceeded 24% in the Florida panther gene pool, and Maehr and Lacy (2002) denounced claims of successful restoration as premature. They suggested that progressive gene swamping might eventually eliminate the Florida genome. They also showed that the contributions of immigrants had been highly unequal. One Texas cougar produced a very fit son who sired at least half of the hybrid cats alive in June 2001.

8.3. Outbreeding Depression

Rather than being too successful, immigrants to a small population can also cause damage if they are not successful enough. This problem, termed *outbreeding depression* or hybrid breakdown (Templeton 1986), happens when the offspring of crosses between populations show lower survival or reproductive success than the original populations. Outbreeding depression commonly occurs when individuals carry genes that are not well adapted to local environmental conditions. Outbreeding depression can also arise when mixing of two gene pools breaks apart pairs or sets of alleles that are coadapted (e.g., Burton et al. 1999). Two alleles are coadapted if the selective advantage of an allele for one gene depends on interaction with an allele for a gene at a different locus. Alleles are *intrinsically* coadapted if the compatibility is independent of the environment (Edmands 2002). The breakup of coadapted alleles does not occur immediately because first-generation hybrids (F_1s) carry a haploid set of chromosomes from each parental line. Segregation and recombination only begin to break apart pairs or sets of alleles in the second (F_2) generation (p. 223 in Lynch and Walsh 1998).

An example of an intrinsically coadapted gene complex occurs in platyfish, a small fish familiar to aquarists. These fish carry a gene, *Tu*, that makes them vulnerable to melanomas, but most individuals also carry a masking tumor-suppressor gene, *R*. When tumor-free fish carrying the tumor gene and the suppressor (genotype *Tu/Tu*, *R/R*) are crossed with a strain that carries neither of these alleles (genotype $-/-$, $-/-$), the F_1s (genotype *Tu/$-$*, *R/$-$*) are all tumor-free because they all carry the suppressor (Adam et al. 1993). Crosses among F_1s, however, yield some F_2s that have extensive tumors. These unfit individuals have an allele for melanoma formation at the tumor-forming locus but no tumor-suppressor allele (genotype *Tu/$-$*, $-/-$ or genotype *Tu/Tu*, $-/-$).

A recent study of freshwater shrimp shows how outbreeding depression can have unfortunate consequences in a conservation program. When shrimp were translocated between pools from two catchments in an Australian river, all resident genotypes were extirpated from one pool after only 7 years (Hughes et al. 2003). The extirpation occurred in two steps. First, translocated males were preferred as mates by both resident and translocated females; second, crosses between translocated males and resident females produced unfit offspring.

8.4. Guidelines for Managing Gene Flow

Because increasing gene flow to a small population is risky, managers contemplating translocations must plan carefully. Six recommendations exist here: (1) Gene pools should be mixed only if significant inbreeding depression has been demonstrated in the target population (Edmands 2002), (2) source and target populations should live in similar habitats and show similar adaptive traits (Storfer 1999; Edmands 2002), (3) tests of the consequences of hybridization should be conducted before any mixing (Edmands 2002), (4) translocated individuals should be marked and genotyped and their genetic contributions to future generations should be assessed (Marr et al. 2002; Stockwell et al. 2003), (5) the number of migrants to a population should be set at 1–10 per generation (Mills and Allendorf 1996), and (6) corrective action should be planned in the event that gene flow exceeds target levels (Maehr and Lacy 2002).

While these guidelines sound reasonable, managers of small populations still face a vital and unanswered question: How likely are genetic restoration and population recovery if these guidelines are followed? Answering this question is difficult because few studies have measured how often gene flow from immigrants increases beneficial genetic variation and population fitness, or reduces population fitness via gene swamping, loss of local adaptation, or genetic incompatibilities.

8.5. Genetic Consequences of Immigration to Mandarte Island

Studies of immigrants to natural populations are rare because they are logistically challenging to conduct. Until recently, immigrants could only be detected if all nonimmigrants were marked before the arrival of immigrants, or if immigrants themselves could be identified before arriving. It is now, however, possible to identify immigrant individuals with molecular markers by using multilocus genotyping or parentage assignment (e.g., Rannala and Mountain 1997; Telfer et al. 2003).

The rich data set on Mandarte Island song sparrows is well suited for studying the effect of immigrants on the genetics of a small population (see chapters 2, 3, 7). Since 1975, the breeding success and survival of all birds on Mandarte Island have been tracked every year except 1980. Thus, immigrants can be identified and a detailed pedigree constructed. In chapters 4 and 7, we learned that an average of 2.8 immigrants join the Mandarte Island song sparrow population per generation. After the population bottleneck in 1989, gene flow from immigrants prevented inbreeding levels from rising for several generations and re-established lost genetic variation. However, immigrant genes may also have prevented purging of the deleterious alleles that cause inbreeding depression (see chapter 7). In the remainder of this chapter, I explore additional ways that immigrants influence the genetics of the Mandarte Island song sparrow population. I first review a recent study on the fitness of immigrant lineages (section 8.6) and then present several new analyses (sections 8.7–8.9).

8.6. Heterosis and Outbreeding Depression in Immigrant Descendants

When Lukas Keller, Peter Arcese, and I studied the performance of immigrant and native song sparrows, we expected the offspring of immigrants to outperform their contemporaries because they would be outbred (Marr et al. 2002). We also expected reduced fitness gains in subsequent generations because heterosis is maximized in F_1 crosses between populations and should decline thereafter (Lynch and Walsh 1998). We considered the possibility that F_1 or F_2 descendants of immigrants might experience outbreeding depression but thought it unlikely. Although the origin of only one immigrant to Mandarte Island was known, short-range dispersal by juveniles occurs frequently among nearby islands (Smith et al. 1996; P. Arcese, A. B. Marr, and S. Wilson, unpublished data). These data suggested that most dispersers moved between islands with similar habitat and environmental conditions. Therefore, differences in adaptive traits seemed improbable. We also believed that gene flow levels were too high for intrinsically co-adapted gene complexes to differ among populations.

To test our predictions, we distinguished five frequently occurring pedigree groups: (1) immigrants, birds that did not hatch on Mandarte Island; (2) natives, birds with two resident-hatched parents and four resident-hatched grandparents; (3) F_1s, with an immigrant parent from either their maternal or paternal side and the other parent and its parents both resident-hatched individuals; (4) F_2s, individuals with an immigrant grandparent and resident-hatched grandparent on both their maternal and paternal sides; and (5) resident backcrosses, individuals with one immigrant grandparent and three resident-hatched grandparents. For these five pedigree groups, we conducted analyses of fitness.

As predicted, F_1s generally performed well. For example, adult F_1s reared as many or more offspring to independence from parental care over their life spans as the average for natives and immigrants (figure 8.1a,b). The survival of juvenile F_1s in their first year also exceeded that of juvenile natives (figure 8.1c). However, in both adults and juveniles, the gains experienced by F_1s were not seen in F_2s. For example, F_2 females reared only half as many offspring as F_1 females, and F_2 males reared only about one-third as many offspring as F_1 males. Survival of F_2 juveniles was about half the survival of F_1 juveniles. On average, F_2 juveniles also showed lower survival than did native juveniles ($p = 0.06$; figure 8.1a), and F_2s bred unsuccessfully compared to adult immigrants and natives of the same sex (figure 8.1b,c). However, although differences in breeding success between F_2s, immigrants, and natives were substantial, they were not statistically significant.

Although we had expected F_2s to perform somewhat worse than F_1s, the large differences here were surprising. This result suggested to us that F_1s benefited from heterosis, but F_2s were probably suffering from a breakup of intrinsically coadapted gene complexes. This was an important finding at the time of publication because it showed that populations of mobile animals could be genetically differentiated at fine spatial scales. Recently, work on great tits demonstrated large adaptive differences in clutch size between populations that exchanged dispersers and were only a few kilometers apart (Postma and van Noordwijk 2005).

Evidence for poor fitness in F_2 song sparrows also led me to question how much immigrant genes affect the Mandarte Island song sparrow gene pool over the long term. Because not all immigrants produced breeding recruits and the fitness of F_2s was poor, immigrant genes could have remained confined to a small portion of the pedigree and been lost within a few years. Alternatively, immigrant genes could have spread widely, diluting or replacing the genes of their predecessors.

8.7. The Spread of Immigrant Lineages and Gene Pool Turnover

To assess the long-term contributions of immigrants to the Mandarte Island song sparrow gene pool, I needed to track the lines of descent through the pedigree for more generations than in Marr et al. (2002). Therefore, I assumed that if a bird had one immigrant parent and one native parent, then half of its genome came from the immigrant parent. If a native bred with a bird that was half immigrant and half na-

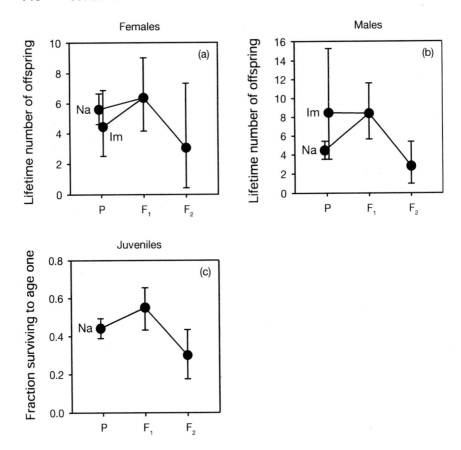

tive, I assumed that the genomes of their offspring were one-quarter immigrant, and so forth. I then summed the contributions of each immigrant to its descendants and divided by the population size to estimate the percentage of the population gene pool that could be attributed to each immigrant in each year since 1981. If an immigrant died without rearing offspring that recruited, then the immigrant contributed to the gene pool during its lifetime but not beyond. I studied only those immigrants arriving after 1981, because many lineages could not be tracked through the gap in banding nestlings in 1980.

One problem with this analysis is that female song sparrows sometimes mate with males other than their social mates (chapter 6), and this causes error in the pedigree (Marr et al. in press). Ideally, we would have used molecular techniques to eliminate these errors, but no DNA samples were collected early in the study, and genotyping of individuals hatched after 1996 remains incomplete. I therefore left any sus-

Figure 8.1. Lifetime reproductive success (LRS) of breeding male and female song sparrows and survival of juveniles from day 24 to age 1. Points show the mean values ± 95% confidence intervals for the two parental groups (P), natives (Na) and immigrants (Im), and for first-generation (F_1) and second-generation (F_2) offspring. Asymmetry in the 95% confidence intervals for the LRS data occurs because the raw data were transformed and the results back-transformed for presentation. Data are from tables 3 and 5 of Marr et al. (2002). Details of statistical methods are given in Marr et al. (2002). In brief, LRS was the total number of offspring that a bird reared to 24 days of age. This analysis included only birds that laid or fathered eggs during their lifetimes. The mean number of offspring raised in each pedigree group was obtained by using the least-square mean estimates from an ANOVA adjusted for year of hatch (a categorical variable). It was necessary to account for year effects because LRS depends strongly on population density, the presence of cowbirds, weather variables, and perhaps other environmental factors in the year of recruitment (see chapter 9). The analysis of juvenile survival to age 1 used a discrete-time proportional-hazards model to compare survival rates by pedigree group. The analysis accounted for year of hatch (a categorical variable) and lay date (a linear variable) in the season. Planned comparisons between pedigree groups were then assessed by t-tests of the least-square mean estimates. The definition of LRS used in this chapter (number of day 24 offspring) differs from that used in chapter 9 (number of recruits). The first definition makes sense here because we are interested in the contributions of parents to offspring survival. These contributions and their biological signal should wane after offspring leave their parents' care.

◄───

pected errors unchanged and proceeded with analyses because my method for studying immigrant contributions is novel and might be helpful to others. Where groups of individuals were compared, misclassification of individuals is likely to have reduced differences among groups, making our statistical analyses conservative (Marr et al. in press).

Many immigrant genes spread widely throughout the population. Four of eight male immigrants and 11 of 17 female immigrants reared at least one offspring that later recruited. One of five male immigrants and 7 of 12 female immigrants had descendants in the population a decade later. The three most successful female immigrants that arrived before 1993 appeared in the pedigree of every new recruit to the population in 2002. One of these three immigrants was wo.gm, the bird introduced at the beginning of this chapter. In addition, the four females that arrived in the 2 years after the bottleneck made a particularly large contribution. In 2002, 4–6% of the gene pool of breeders

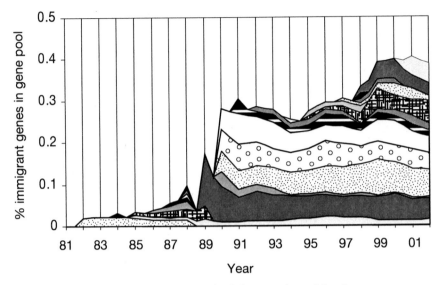

Figure 8.2. Fraction of the gene pool of the Mandarte Island song sparrow population that can be traced to immigrants arriving after 1981. The contribution of each of 25 immigrants is depicted with a different shading pattern. Some contributions are small and barely noticeable.

stemmed from each of these four immigrant females (figure 8.2). The male immigrant that arrived in 1990 also contributed to the gene pool, but only by a meager 0.4% in 2002.

By 2002, 39% of the gene pool of breeding birds could be traced to immigrants that joined the population after 1981 (figure 8.2). The remaining 61% of genes stemmed from 21 of the 47 birds that were present in 1981 and from three resident-hatched birds of unknown parentage that fledged unbanded in 1992. Taken together, these analyses suggest that immigrants made a substantial contribution to the population gene pool over a 21-year period.

8.8. The Competitiveness of Immigrant Genes

To further understand the relationship between immigrants and gene flow, I also tested if immigrant genes were outcompeting native Mandarte genes or vice versa. Based on the analyses of lifetime fitness described above (section 8.6; figure 8.1), I hypothesized that immigrant genes would first spread quickly relative to native genes and more slowly thereafter. This pattern is expected because F_1s survive well to age 1 and are fecund as adults, but F_2s perform poorly. However, the spread of immigrant genes also depends strongly on the relative fitness of back-crosses and their descendants in later generations. In addition, in species that breed in multiple years, the rate at which an individual's genes spread will depend on when in life it breeds, the breeding age of its offspring, and so forth. I therefore compared the relative contributions of individual immigrants to natives of the same sex from the same cohort, using the method for estimating relative contributions to the gene pool (see section 8.7).

Female immigrants contributed more to the gene pool of the breeding population than did natives 1 year after their arrival, but this advantage diminished just 2 years after their arrival (figure 8.3a). Eight years after their arrival, the genes of female immigrants were 25% less common on average than the genes of native females from their cohort. This difference, however, was not statistically significant (figure 8.3a). For males, immigrant genes spread erratically, probably due to the small sample sizes of immigrant males that bred ($n = 4$ breeding immigrants; figure 8.3b). However, over the long term, immigrant males were as successful as native males.

8.9. The Effect of Immigrants on Neutral Genetic Variation

In sections 8.7 and 8.8, I examined gene flow from immigrants by estimating the proportional arithmetical contributions of immigrants and residents to the population pedigree. Another way to study the genetic contributions of immigrants is through allelic diversity at neutral marker loci (e.g., Grant et al. 2001; Keller et al. 2001). In particular, I asked if immigrants altered genetic variation in the population gene pool. For this analysis, I used all genotypes available for adult birds from 1987 to 1996 at eight microsatellite loci. Blood sampling effort varied across years, ranging from 45% to 100% of the population (table 8.1).

Immigrants restored some alleles lost during the 1989 bottleneck over the next 7 years. In 1987, genotypes were available for 92 resident adults (73% of all adults alive; table 8.1). Together, these birds carried 60 alleles at the eight loci (figure 8.4). Ten years later, 12 of the 60 alleles (20%) were extirpated from the population, nine others (15%) had been extirpated and reintroduced, and 39 of 60 (65%) were present continuously (figure 8.4). Over this same 10-year period, 21 new alleles were detected (figure 8.4). Two of these 21 new alleles were

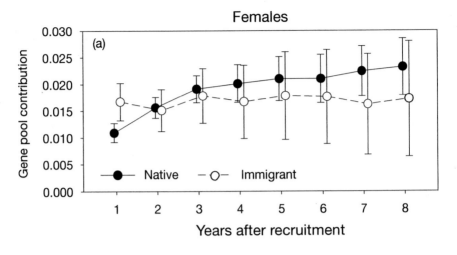

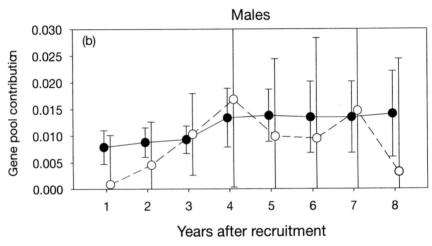

Figure 8.3. The contributions of individual female and male song sparrows to the Mandarte Island gene pool over time. Points show the fraction of the gene pool of the population attributed to the average immigrant individual (open circles) and native individual (solid circles) ± 95% confidence intervals in the 8 years following their recruitment (see section 8.8). The contributions of immigrant and native birds to the gene pool were compared using general linear models that first controlled for year effects. Immigrants introduced more genes to the gene pool than native females 1 year after they arrived ($p = 0.004$; $n = 16$ immigrants, 87 natives), but there were no significant differences between females in subsequent years (all $p > 0.25$). Note, however, that there were only 16 immigrant females in this analysis, making interpretation of small differences difficult. The contribution of male immigrants and male natives to the gene pool did not differ significantly one year after they arrived ($p = 0.16$; $n = 4$ immigrants, 38 natives) or in any subsequent years (all $p > 0.33$). However, only four immigrant males ever bred.

Table 8.1. Numbers of adult birds in the Mandarte Island population and the percentage that were genotyped in each year from 1987 through 1996.

Year	Number of birds	% Genotyped
1987	126	73
1988	105	84
1989	12	92
1990	24	63
1991	55	45
1992	87	72
1993	113	93
1994	127	99
1995	112	99
1996	128	100

Note. In 1989, the year that was most crucial to analyses because of small population size, samples were available for 11 of 12 birds. The bird that was not sampled was an unmated male that never bred.

extirpated a year after being introduced. Thus, by 1996, 128 adults carried 67 alleles at the eight loci, and 19 of the 67 alleles (28%) were not present in the 2 years before the bottleneck.

   The contribution of individual immigrants also varied strongly over time. The two genotyped immigrants that arrived in 1987 and 1988, when the breeding population was large (≥ 105 individuals), brought no unique alleles with them. The three genotyped immigrants that arrived in the 2 years after the bottleneck in February 1989 brought eight, five, and four new alleles each. Four immigrants that arrived from 1991 to 1995 each carried one to three new alleles.

8.10. Synthesis and Recommendations for Future Study

Understanding the effect of immigrants on gene flow and fitness in small populations is important to the management of rare species. In this chapter, I have described the effect of natural immigrants on the genetics of one small population. Immigrant song sparrows to Mandarte Island arrived at a trickle of about three birds per generation (see chapter 4), a level that is within the range recommended for maintaining genetic variation in small populations (see Mills and Allendorf 1996). At this level of immigration, inbreeding still occurred regularly and in-

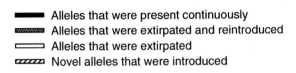

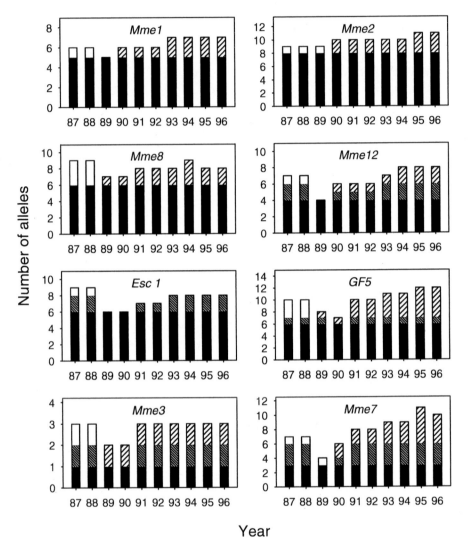

Figure 8.4. Count of alleles present in the population at each of eight microsatellite loci from 1987 through 1996. The names of the microsatellite loci are listed at the top of each graph. The population experienced a severe bottleneck in 1989.

breeding depression lowered the fitness of some individuals in most years. Yet, the immigrants made a valuable genetic contribution. Although extrapair fertilizations must have created errors in the pedigree, these analyses suggest that many immigrants brought new genes and the genes of some individuals spread throughout the population. Together, the contributions of immigrants caused substantial gene pool turnover. By 2002, about 39% of the gene pool could be attributed to immigrants that joined the population after 1981. Analysis of neutral molecular markers showed that immigrants reintroduced some genetic variation that was lost during a population bottleneck as well as many other new alleles.

Interestingly, although the Mandarte song sparrow population is inbred, only the F_1 descendants of immigrants experienced heterosis, and immigrant genes did not spread faster than native genes in the long-term. These results for song sparrows contrast with results on squinting bush brown butterflies. In the butterflies, heterosis accelerated the spread of immigrant genes to inbred laboratory populations over several generations (Saccheri and Brakefield 2002; see section 8.1). Perhaps this difference occurred because the laboratory butterfly populations were more inbred than typically occurs in nature (Gaggiotti 2003). Locally adapted genes and intrinsically coadapted gene complexes may have also had less effect on fitness in the benign conditions of a laboratory. Alternatively, heterosis lasting for multiple generations may be the usual pattern in populations undergoing genetic rescue.

Future studies aimed at predicting patterns of genetic rescue would be valuable, including descriptive research on natural populations. Experimental translocations in the wild are open to ethical objections, but much could be learned from improved monitoring of translocations made for conservation purposes. Few studies to date have tracked the contribution of released individuals or natural immigrants beyond the F_1 generation, but documenting the fitness of F_2s and F_3s is at least as important. Additional laboratory studies of patterns of heterosis, gene swamping, and outbreeding depression could also be rewarding. It would be particularly helpful to see more studies that mix gene pools of populations that have been inbred to varying degrees, or isolated for varying periods.

9 Unequal Lifetime Reproductive Success and Its Implication for Small Isolated Populations

Wesley M. Hochachka

Male song sparrow rwo.mo (banded red-white-orange on the left leg, metal-orange on the right) was born in 1981 to two yearling parents. He settled near the center of Mandarte Island while the population was recovering from the crash of

1979–1980 and remained on the same territory until his death. Over his 7-year life, he left eight breeding offspring from 25 nesting attempts. While his longevity allowed him to sire many recruits (most breeders from his cohort left only 0–2 recruiting offspring), he fell well short of his sister's efficiency. She required only eight attempts over 4 years to leave nine recruiting offspring. While longevity increases the likelihood of high lifetime reproductive success, there is still a large element of chance. Male rwo.mo died in the fall of 1988, eaten by a passing northern shrike that arrived on Mandarte and left the same day.

The number of offspring that an individual produces in its lifetime, its lifetime reproductive success (LRS), has been of interest to many evolutionary ecologists as a measure of an individual's realized fitness (Newton 1989b; Partridge 1989). Much of the original interest in measuring LRS stemmed from a desire to identify selection pressures on organisms (Clutton-Brock 1988; Newton 1989b). However, various studies have shown that LRS is only a true measure of fitness under a relatively narrow set of conditions, in particular, when populations are constant in size (e.g., Benton and Grant 2000; Brommer et al. 2002). Even stochastic variation in population size reduces the value of LRS as an approximation of fitness, because young born early and late in a parent's life may differ in value when a population is increasing or decreasing in size (Cole 1954). The search continues for a robust empirical measure of fitness, because all currently proposed measures have some weaknesses (Brommer et al. 2002; Link et al. 2002).

Regardless of the utility of LRS as a precise measure of fitness, examining variation in LRS can illuminate both evolutionary and ecological problems. For example, variation in reproductive success and, in particular, skewed distributions of reproductive success among individuals are central to the evolution of social grouping in birds (Magrath and Heinsohn 2000; Haydock and Koenig 2002), other vertebrates (Cooney and Bennett 2000; Balshine et al. 2001), and social insects (e.g., Bourke and Heinze 1994). For this book, the importance of variation in LRS is that it reduces the effective size of populations and hence can affect the genetic integrity of small and isolated populations (see chapter 5).

LRS has varied greatly among individuals on Mandarte Island. Females have produced between 0 and 18 breeding offspring, while male

LRS has varied from 0 to 23 breeding offspring. We already have considerable information about the causes and consequences of this variation, thanks to two previous examinations of the data (Smith 1988; Hochachka et al. 1989). This previous work has shown that the populationwide distribution of LRS is highly skewed, with most adults leaving few or no offspring, and a very few contributing disproportionately to subsequent generations (Smith 1988). A similar strongly skewed pattern of variation in LRS is typical of most other species for which estimates are available (Clutton-Brock 1988; Newton 1989a).

The main causes of variation in LRS of song sparrows on Mandarte Island are variation in adult longevity and juvenile overwinter survival (Smith 1988), with potential recruits being less likely to enter the breeding population at high population densities (Arcese et al. 1992; see chapters 4, 6). Thus, LRS of adults is also a function of the population density in the year in which they started nesting (Hochachka et al. 1989). We have also shown that, at least in some years, not just who a bird is (Smith 1981b) but where that bird lives on Mandarte can affect annual and lifetime productivity (Hochachka et al. 1989).

What has not been done up to now, however, is to explicitly place these facts within the context of the dynamics of a small and relatively isolated population and ask whether the above patterns have consequences for the stability of such populations. Small populations may decline because of catastrophe, genetic problems, or demographic stochasticity (Shaffer 1981; see chapter 1). The effects of catastrophes are covered elsewhere in this book (see chapters 4, 10), as are the problems caused by the accelerated loss of genetic diversity that results from catastrophic population declines (see chapters 1, 7). However, even without population crashes, LRS has been unequal among adults, and the rate of loss of genetic diversity has thus been higher than it would have been with more equitable reproductive success.

My overall goal in this chapter is to identify factors that cause one group of birds to have higher reproductive success than another, thus increasing skew in LRS. In particular, I explored whether the demographic problems due to decreased genetic diversity and lower reproductive rates are likely to occur under the same conditions, potentially compounding two risks of increased population extinction. In conducting many analyses, I have excluded the data[1] that come from the population crashes in 1979–1980 and 1988–1989, because inclusion of these data can affect the patterns produced from analyses and thus possibly limit the generality of any conclusions to other populations or species. Nevertheless, the effects of population crashes are represented indirectly through the effects of these events on variation in population density. The data from population crashes were included in our previous analyses of variation in LRS (Smith 1988; Hochachka et al. 1989).

One part of addressing my overall goal is to conduct more detailed analyses to test for subtler or more complex effects on LRS than were examined previously. For example, I will test whether the lower survival of later-hatched offspring (Hochachka 1990) is exaggerated in years with low overwinter survival of juveniles (Arcese et al. 1992), and whether there are subtle effects of population density that act only on adults of specific ages. Some of the analyses parallel work already discussed in chapters 3–5. In this chapter, the analyses were redone in order to allow comparison of all potentially relevant influences on adult and juvenile survival within a single location, to minimize the need for readers to jump between chapters of this book.

Another part of addressing my main goal will be to revisit earlier analyses and conclusions. Previous analyses of patterns in LRS of Mandarte's song sparrows were conducted on data from relatively few cohorts of birds, five cohorts in Smith (1988) and nine in Hochachka et al. (1989). Elsewhere in this book are examples where prior patterns and interpretations have not been upheld; contrast the results in chapter 4 with Arcese et al. (1992), and chapter 7 with Keller (1998). The same is true here, where I reassess the previous conclusion (Hochachka et al. 1989) that certain locations on Mandarte Island were better than others for song sparrow reproduction, an environmental cause of increased skew in LRS.

9.1. Causes of Variation in Adult Survival

Previous analyses (Smith 1988) have shown that the variation in LRS among adults was largely explained by how long birds live. Thus, in order to understand the causes of varying LRS, we would like to identify factors that affect the probability of adult survival. In this section, I describe relationships between overwinter survival of breeding adults and several intrinsic and extrinsic factors. In particular, I have re-examined the findings that adults that produce more offspring survive better (Smith 1981b) and that adult overwinter survival is not affected by population density (Arcese et al. 1992). In addition to looking for overall effects of reproductive success and density, I also tested for variation in effects between sexes and among adults of different ages, because of the effects of skewed sex ratios on population growth rates (see chapter 4) and the importance of adult longevity to LRS (Smith 1988). Given that annual reproductive success of adults is an index of individual quality, and given that quality might have a genetic basis (see chapter 7), I also examined whether siblings have similar survival probabilities. The effect of cowbird parasitism on survival of adult song sparrows has been described in chapter 5, but here I revisit this using a different analytical framework (Burnham and Anderson 2002), which acknowledges that factors may have biologically real, although statistically nonsignificant, effects (Johnson 1999).

Factors Affecting Adult Survival

Some intrinsic characteristics of adult song sparrows affected their probability of overwinter survival.[2] This analysis showed, as expected, that an adult's age affects the probability that it will survive to breed again (figure 9.1). Older song sparrows survived less well on average. Males had slightly higher survival probabilities than did females, although survival rates were statistically indistinguishable (overlapping confidence

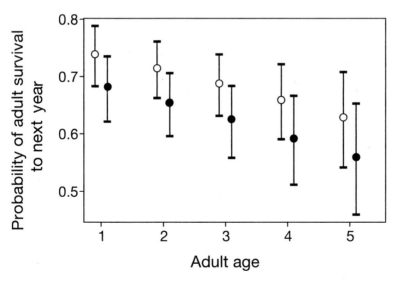

Figure 9.1. Probability that an adult song sparrow will survive to the next breeding season, as a function of its current age. Plotted values are model-averaged estimates of survival probability (see note 2). Solid circles indicate predicted survival probabilities for females, and open circles are estimated survival probabilities for males. Error bars indicate 95% confidence limits.

limits). However, I was not able to reproduce the earlier result (Smith 1981b) that adults raising a larger number of offspring also had a higher probability of overwinter survival.

After accounting for age-related differences in survival, I found that there were still consistent intrinsic differences in survival probability not just among individual birds but among entire groups of siblings. This conclusion was based on examining the variation in survival probability that could not be predicted by any specific individual trait or environmental characteristic. Roughly 9% of this "random" variance in survival was accounted for by similarity in the likelihood of overwinter survival of siblings born in the same year. By comparison, interannual variation in survival rates of adults accounted for 12% of the same total "random" variance. Thus, differences among families are a lesser source of variation in adult survival than are year-to-year differences. Nevertheless, it is notable that any repeatability in survival within families was detected.[2]

Both of the extrinsic factors I tested, population density and presence of cowbirds, had little effect on adult survival. While my analyses suggested that there was some effect of population density on adult survival, the estimated annual survival probabilities only varied by 0.005% between the years of lowest and highest density. Likewise, while statistically supported, the effect of cowbird presence on adult survival only accounted for a predicted difference in annual survival of 0.003% between cowbird and noncowbird years. This conclusion held true whether I tested for identical effects of density on all adults (Arcese et al. 1992) or looked for effects of density that differed with adult age or that affected only adults of one sex and age.[2] If anything, the trends here were in the opposite directions of those expected: Adults survived slightly better at high densities and in years with cowbirds.

Implications

As in previous analyses, few factors affected survival of breeding song sparrows. The decline in survival with advancing age (figure 9.1) was not altered by population density, nor did it vary substantially between sexes. The stress of coping with cowbird parasitism had no detectable effect on adult survival, nor was there any indication that the number of independent offspring reared was related to parents' survival probability. The most interesting finding was that adult siblings shared similar likelihoods of overwinter survival. One cause of such a pattern might be genetic (see chapters 7, 8). Alternatively, this pattern could be environmental in origin, the result of variation in developmental conditions among nests or territories; "nest" or "territory" effects could not be distinguished in these analyses. Aside from the differences in adult survival with age and among sibling groups, my analyses show that while adult survival has a substantial influence on LRS, it is mainly chance that explains why some adults have short lives and others long ones.

9.2. Causes of Variation in Overwinter Survival of Juvenile Song Sparrows

The second most important source of variation in LRS of song sparrows on Mandarte was variation in overwinter survival of offspring after they became independent of their parents (Smith 1988). We have previously examined several factors that could affect survival of offspring (Nol and Smith 1987; Hochachka 1990; Arcese et al. 1992), but there is still scope for further analyses. New patterns might be revealed with the larger sets of data now available. For example, my detection of reduced adult survival with increasing age (figure 9.1) was not found with a subset of these data (Nol and Smith 1987). Thus, I reappraised the roles of nestling hatching date (Hochachka 1990) and brood size/nestling condition (Hochachka and Smith 1991) on juvenile survival. I also examined whether the detrimental effects of late hatching are compounded by being born into a larger brood or in a year of high population density. Lower survival of juveniles at higher population densities is examined more closely, in order to determine whether interactions among juveniles, or between juveniles and adults, are at the heart of the known effects of population density (Arcese et al. 1992). Finally, I extended the examination of relationships between adult age and nesting success (Nol and Smith 1987) by formally testing whether adult age affected overwinter survival of independent offspring (see figure 3.4 for a graphical examination of the same issue).

Many factors can affect overwinter survival of juveniles. I selected factors here that I thought might accentuate disparities in LRS among adult song sparrows. Changes in offspring survival with increasing adult age could either compound or lessen the disparities in LRS stemming from variation in adult longevity, if older adults produce offspring with higher or lower probabilities of survival. In a similar vein, the lower survival of later hatched offspring could compound differences in LRS among adults, because the youngest adults generally nest later (see figure 3.4). I quantify the relative sizes of these effects below.

Factors Affecting Overwinter Survival of Juveniles

My analyses[3] indicated that hatching date, father's age, adult density, and juvenile density all affected overwinter survival of juvenile song sparrows. Of these, date of hatching had the largest effect (figure 9.2): Juveniles from a year's earliest nests were twice as likely to survive the winter on Mandarte, compared with the last hatched juveniles. Variation in the density of breeding adults also had a large impact on juvenile survival (figure 9.3), with survival at low adult density being up to twice that in high-density years.

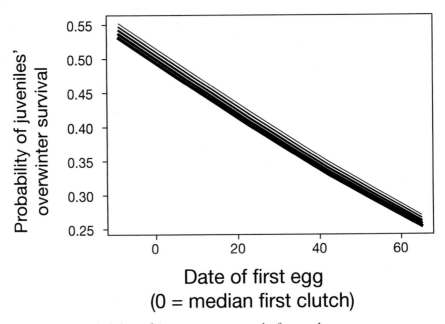

Figure 9.2. Probability of over-winter survival of juvenile song sparrows as a function of hatching date (estimated by the laying date of the first egg in the clutch), and father's age (see note 3). Predicted values were calculated for an adult density of 42 breeding females and a juvenile density of 123 independent juveniles. Date zero is the median date of first nests for each year, and the range of laying dates represents the 5% and 95% quantiles for standardized laying date. Ninety-five percent confidence limits extended roughly 0.1 above and below the predicted values; thus, survival probabilities of juveniles from different-aged fathers were statistically inseparable. The thinnest (upper) line is the predicted probability of survival for the offspring of 1-year-old fathers, and progressively thicker lines are predicted values for fathers' ages 2 through 5.

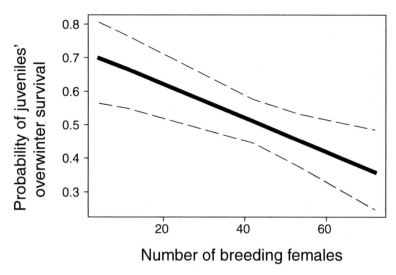

Figure 9.3. Probability of overwinter survival of juvenile song sparrows varied with changes in adult density. Adult density is the number of breeding females in the year of juveniles' birth. Probabilities (solid line) and associated 95% confidence limits (dashed lines) are based on a father's age of 2 years, at the median date of first nest (standardized date = 0), and 123 independent juveniles.

In contrast to these strong and statistically well-supported effects, the effects of juvenile density and fathers' age on survival were both small and only weakly supported. Older fathers tended to produce offspring with a lower survival probability (figure 9.2). Juveniles tended to survive less well in years of high juvenile densities, although estimated juvenile survival varied by less than 2% between the years of highest, and lowest juvenile density and confidence intervals around this estimated effect broadly overlapped zero. Juvenile and adult densities were partially confounded, because more juveniles were produced in years of high adult density. However, the correlation was low enough $(r^2 = 0.34)$ that the majority of variance in juvenile density was unexplained by variation in adult density. Thus, the low impact of juvenile density on juvenile survival is probably biologically real and not a statistical artifact of collinearity between the two density variables.

I also found that siblings born in the same year tended to have similar probabilities of overwinter survival. Roughly 12–13% of all "random" variance not accounted for by the predictors described above was due to variation among sibling groups. By comparison, interannual variation was roughly 17–18% of all random variance.[3]

Implications

The two strongest influences on overwinter survival of juveniles both contributed to skewed distributions of LRS of song sparrows on Mandarte Island. Variation in LRS among cohorts was amplified by variation in survival probability with adult density (figure 9.3) because of the relatively short life spans of breeding sparrows and similarities in population densities between successive years (Hochachka et al. 1989). Within cohorts of adults, early breeding by older sparrows (Nol and Smith 1987) and declines in survival with later nesting date (figure 9.2) combine to accentuate the differences in LRS between sparrows that nest for a single year, relative to adults that breed in 2 or more years. Additionally, the observed variation in survival probability among sibling groups will also contribute to increased skew in LRS, both within and possibly among cohorts. However, my analyses[3] suggest that only siblings born in the same year tended to have similar fates, meaning that any given adult will not tend to produce high- and low-survival offspring throughout its life.

Given that Mandarte is part of a larger metapopulation of song sparrows and that my measure of juvenile survival is local survival, the full implications of variation in juvenile survival are unknown. Some of the juvenile birds that failed to survive on Mandarte emigrated, and some of these undoubtedly bred nearby. Thus, particularly in the aftermath of population crashes, some of these emigrants' genes may be reintroduced onto Mandarte and spread widely in the population (see chapter 8) generations after the emigrant was discounted as part of its parents' Mandarte-specific LRS. If the rate of juvenile emigration was higher at high population densities or for later-hatching juveniles, fitness measured over many generations or over the larger meta-population may differ substantially from that viewed over a few years on Mandarte.

9.3. Spatial Variation in the Likelihood of Nest Failure

I have noted above how conditions external to a nesting song sparrow can affect its LRS, with lower offspring survival and hence LRS for sparrows breeding in years of high sparrow density (figure 9.3). All breeding sparrows experienced this effect of high density about equally.

I did not find strong differences in the effect of density on survival probability of offspring among different groups of breeding adults; neither adult age (figure 9.2) nor nesting date[3] affected the survival rates of offspring differently in years of high or low population density. However, variation in territory quality may accentuate differences in LRS of breeding adults. Hochachka et al. (1989) found that some locations on Mandarte Island tended to consistently experience higher probabilities of nest failure, although these "poor" locations were scattered throughout Mandarte. Given the high site fidelity of breeding song sparrows (see chapters 2, 6), variation in territory quality could produce some of the observed differences in LRS.

The conclusions from our previous analyses of the determinants of LRS (Hochachka et al. 1989) are weakened by the fact that year-to-year variation in nesting conditions was only crudely controlled by restricting analyses to an arbitrary set of years with similar nesting density. A more detailed examination, which controls for other factors affecting the probability of nest failure, is the subject of the following analyses.

Do Nests Fail Repeatedly in Some Locations?

The idea behind this analysis is that, after other sources of variation in nesting success are accounted for statistically, similarities in the probability of nearby nests failing would still exist if there were good and bad locations for nesting on Mandarte. These similarities would result in highly correlated residual variation in success in nearby nests. In my analyses,[4] I created a statistical model describing the probability of nest failure based on known or suspected influences on nest failure. The residuals from this analysis were then examined for declining similarity in the unexplained variation in survival as distances between nests increased. The expected pattern here is of greater dissimilarity in failure probabilities as distances between nests increased.

I found evidence for "poor location" effects only for nests produced in the same year (figure 9.4a). There was no suggestion that areas were consistently poor even in 2 successive years (figure 9.4c). Even within the same year, any shared chance of failure disappeared at internest distances beyond roughly 5 m.

Nests within 5 m of each other—5 m was the distance at which the level of dissimilarity in fate reached a maximum (figure 9.4a)—were mostly produced by the same female, making it impossible to distinguish between effects of location per se on nesting success and ef-

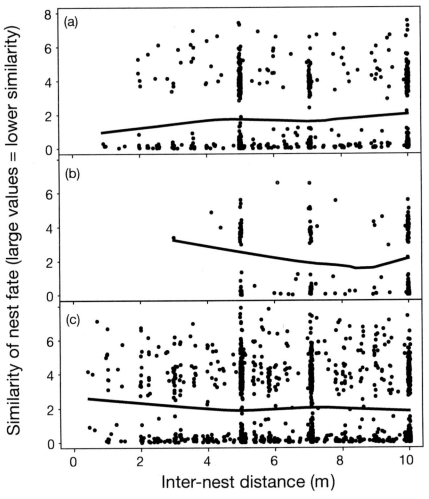

Figure 9.4. Exploration of whether probabilities of nest failure vary in space across Mandarte Island. Dissimilarities in fates (failed, or produced independent offspring) of pairs of nests are plotted as a function of distances between nests, after controlling for multiple sources of variation in nesting success (see note 4). If nearby nests share similar fates, then dissimilarities should be smallest at closest distances, with dissimilarities increasing to a plateau as inter-nest distances increase. (a) data from nests in the same year. The line through (a) reaches a plateau at internest distances of roughly 5 m, suggesting that nests beyond this distance apart were independent in their fates. (b) The same data, but without pairs of nests that were produced by the same female, to disentangle location-specific effects from combined female + location effects. (c) Data from all pairs of nests that were produced 1 year apart. Lines are LOWESS smoothings. A small random error was added to the x-values for each plotted point to allow visualization of the number of data points at each internest distance.

fects of parents' nest site selection and/or behavior (Smith et al. 1984). When data from pairs of nests of the same parents were removed (figure 9.4b), there was no trend for nearby nests to share the same fate.

Implications

Using all of the available data, I was not able to confirm the presence of site-specific variation in nesting success on Mandarte Island. The only indications of site effects were likely within individual territories, and then only within a single year (figure 9.4). As such, it is impossible to determine whether the site per se or the birds occupying the site were responsible for the observed patterns. Thus, I was not able to confirm the existence of site-specific variation in nesting success. I conclude that variation in territory quality, as it affects nesting success, has contributed little to variation in LRS among the song sparrows on Mandarte Island.

9.4. Discussion

Because of the small size of the song sparrow population on Mandarte Island (see chapters 1, 7), loss of genetic variability is expected to occur quickly due to genetic drift. Skewed distributions of LRS among birds born into the same cohort or among cohorts of birds will reduce the effective population size, N_e, and accelerate the rate at which genetic variability is lost from a population (Crow and Kimura 1970). The goal of this chapter has been to describe factors that accentuate disparities in LRS and thus accelerate the rate of loss of genetic information within a small population. These disparities occurred both among and within cohorts of song sparrows born on Mandarte (Hochachka et al. 1989).

The expected rate of loss of genetic variability would be constant among cohorts, if there were no interannual variation in survival (see chapters 3, 4, 7), nesting success (see chapters 4, 5), or mating system (see chapter 6). On Mandarte, all these traits varied across years, and thus the skew in distribution of LRS has changed through time (Hochachka et al. 1989). Catastrophic population crashes have played a role in creating this variation (Keller et al. 2001; see chapter 7).[1] However, even after the effects of population crashes were removed, there was still variation in the skew of LRS among cohorts of song sparrows and therefore variation in the rate of loss of genetic information from year to year. Given that most variation in LRS comes from variation in adult survival and juvenile survival (Smith 1988), variation in these parameters is probably responsible for much of the observed variation in skew of LRS.

While survival does vary among adults, almost none of this variation could be explained by any of the factors that I examined. This is in keeping with previous examinations of variation in survival of adult song sparrows on Mandarte. Of particular relevance to this chapter, there was no evidence that various factors acted together to increase or decrease skew in LRS. Specifically, I found no indication that overwinter survival of adults was affected by population density or reproductive effort, either for all adults or for a limited set of ages.[2] The only pattern detected in survival probabilities of adults was that adult siblings born in the same year tended to have similar chances of overwinter survival, although this pattern did not hold for siblings born in different years.[2] While similarity in the survival rates of adult siblings could indicate differences in genetic quality, it is more likely that they were effects of a specific rearing environment because similarity in survival was only noted for offspring born in the same year. Alternatively, the patterns here may be caused by genetic effects being expressed differently for different cohorts of siblings either as a function of the environments in which siblings grew up, or of the environments in which they lived as adults. Whatever the case, a simple genetic basis for this pattern is unlikely, because this should cause siblings having similar fates across all years. Poor conditions in the nest have long been known to affect juvenile survival in birds strongly (e.g., De Kogel 1997; Naef-Daenzer et al. 2001). Also, the possibility that poor nestling condition can have longer term consequences for adults (Schluter and Gustafsson 1993) has been receiving increased attention recently (e.g., Metcalfe and Monaghan 2001; Cam et al. 2003). Whatever the cause, similarities in survival probabilities among siblings will accelerate the loss of genetic diversity because some entire genetically related groups will have fewer reproductive attempts due to their shorter lives.

In contrast to adult survival rates, and in keeping with previous analyses of the data from Mandarte Island, variation in juveniles' overwinter survival was predicted by a number of factors. A later date of birth (figure 9.2) and a higher number of adults in the year of birth (figure 9.3) both contributed to lower overwinter survival of juveniles. After such effects were accounted for in my statistical analyses,[3] I still found consistent variation in survival among sibling groups. As with the similar survival of adult siblings, similarities in juvenile survival were found only for offspring born in the same year, suggesting that these patterns did not have a simple genetic basis. Note that while various factors affected juvenile survival, none had complex interactions that altered survival of only subsets of juveniles, aside from seasonal declines in survival probability (figure 9.2). However, I found neither variation in the effect of laying date with population density nor variation in the effect of laying date with variation in brood size.

All of this systematic variation in juvenile survival serves to accelerate the loss of genetic variability at high population densities. Similar survival among siblings obviously increases reproductive skew within cohorts. This effect is compounded by low juvenile survival at high population densities (figure 9.3), which creates differences in average reproductive success among cohorts of breeders. A subtler cause of increased reproductive skew is through variation in juvenile survival with laying date (figure 9.2), because older parents typically nest earlier (see chapter 3). Given that more than 35% of breeders survive to nest only a single year, the effects of variation in adult survival on LRS are compounded by the added differences in average laying date.

If these patterns can be generalized to other small populations, three conclusions can be drawn. First, variation in adult survival, while it is an important source of variation in LRS, showed little systematic variation aside from consistency in survival probability among siblings from a single year. Thus, variation in adult survival rates is principally random fodder for genetic drift and not a source of predictable variation in the rate of loss of genetic information from a population. Second, in contrast to adult survival, survival rates of juveniles will respond to variation in population size (figure 9.3) and thus should serve as a major source of systematic variation in the rate of loss of genetic variation as population sizes vary. Third, the effect of changing population density on reproductive skew, and N_e will be closer to the actual population size at low population densities than at high densities.

Notes

1. I tested for trends in the relationship between statistical skew and the number of years to a crash (figure 9.1) using linear regression analyses (SAS PROC MIXED). I compared models of constant skewness, a linear trend, and a quadratic trend with changes in the duration before a population crash. Support was judged using the bias-corrected version of Akaike's Information Criterion scores (AIC$_C$; Burnham and Anderson 2002), with lower AIC$_C$ score indicating better support for the model. For males, the respective AIC$_C$ scores were 18.1, 23.0, and 23.6, indicating that roughly 87% of all of the support from the data was for the model with no variation in skewness with years to

crash; this conclusion is based on calculating Akaike weights (Burnham and Anderson 2002). For females, the respective AIC_C scores were 192.8, 188.4, and 187.0. The strongest model support for females came for the model with a quadratic shape (lowest skewness at intermediate times to a crash), but with only 64% of all support for this model and a combined support of roughly 96% for models that had a change in skewness of LRS with variation in the number of years to a population crash. Thus, while there is reasonable statistical support for a change in skew of LRS with the number of years to a population crash, none of the three statistical models clearly described the pattern well. The high combined support for models that described some change in skewness was largely caused by the high skew for the cohort that reproduced for only 1 year before the population crash of 1979–1980. The above analyses used data from adult birds that had at least one nesting attempt in their lives. When data from all birds (i.e., including floating or unmated males) were included in my analyses, the results were qualitatively similar. An electronic appendix to this chapter, including a more detailed description of these analyses, is available from the author.

2. Testing for effects on adult survival was carried out in three stages, using the multimodel inference paradigm (Burnham and Anderson 2002). In the first stage, a set of 20 statistical models was created to test for effects of adult age and annual reproductive success (number of independent offspring) on adult survival. In all statistical models, the adults' sex and age (linear continuous variable) were included. The possible effects of two experiments (the 1985 and 1988 food supplementation experiments) were accounted for by including the type of experiment, as a categorical variable, in the statistical models. Mixed-model logistic regressions were performed using the GLIMMIX macro and PROC MIXED of SAS, using maximum likelihood. The dependent variable was binomial; that is, a breeding adult either survived to the beginning of the next breeding season or did not. Two random effects were included in the initial set of analyses: year and a unique family identifier (used to identify full siblings as a group); both were treated as categorical variables with a compound symmetrical covariance structure. A table listing these 20 models and their relative support is available in the digital appendix to this chapter, which is obtainable from the author.

Five statistical models had ΔAIC_C scores <4 from the best model, indicating that any of these models had some likelihood of being the most appropriate description of the causes of variation in adult survival rate. All of the successful models had very few parameters. Two models contained effects of population density on survival, but none contained effects of annual reproductive success. The second step in the analysis process was to take the five best-supported models from the initial set and add presence/absence of cowbirds as an additional explanatory variable to each of these, forming a new set of 10 models (five without cowbird effects, and five including cowbird presence/absence). In this new set, eight models had $\Delta AIC_C < 4$, including three models where the presence or absence of cowbirds affected song sparrows' overwinter survival.

The final step in the analysis was to assess which random effects were best supported by the data. Year was always included as a random variable, based on our prior knowledge that interannual variation in adult survival exists. The random effect that I explored in more detail was similarity in survival probability of full siblings. The two previous steps in this analysis assumed that all full siblings has correlated probabilities of survival, but I explored two other alternatives: (1) There was no similarity in survival among siblings, and (2) similarity in survival rates was of environmental and not genetic origin and therefore only siblings born in the same year (and likely from the same nest) would have similar survival probabilities. I compared the relative support for these three alternative effects of relatedness in the final step in the analysis of adult survival rates. I did this by taking the eight reasonably supported models identified in the second step and creating three statistical models based in each of the eight combinations of fixed effects (a total of 24 models). Each trio of models with the same fixed effects had one model each with the following random effects: (1) year only, (2) year and annual sibling group (only full siblings born in the same year grouped), and (3) year and sibling group (with full siblings across years combined into one group). Analyses were conducted as above, except that restricted maximum likelihood was used in GLIMMIX. For each of the eight different groups of fixed effects, the best-supported random effect was always the annual sibling grouping. That is, siblings born the same year tended to have similar probabilities of overwinter survival, even after interannual variation in survival was accounted for in the statistical models.

The eight best-supported fixed effect models were then combined with the best-supported random effects, and predicted values were calculated. Predicted values and their standard errors were averaged across models and are presented in figure 9.1. "Model averaging," briefly, is the process of producing a value for a regression coefficient or a predicted value when there is no single clearly best model, but when several models all have reasonable support from the data. In such cases, the values produced are weighted averages across statistical models, with weights being the Akaike weights for each of the statistical models. Model averaging of standard errors and other measures of variance is slightly more complicated, because the weighted averages are inflated systematically to take into account not just within-model uncertainty as to the correct value (which variance estimators provide) but also among-model uncertainty of the correct values. A detailed explanation of model averaging is provided in Burnham and Anderson (2002). Averaging was conducted across all well-supported models ($\Delta AIC_C \leq 4$). When a predictor variable was not present in a model, this effectively means that the predictor variable was actually there but its value was set to zero (i.e., the predictor had no influence). These zero-valued predictors were included in the model averaging.

In these analyses, the number of breeding females in the year prior to over-winter survival was used to represent population density. I also explored whether using either the number of adults of the same sex as the bird in question or the total number of breeding adults was a better index of population density. However, neither measure produced a better-supported fit to the data, and all three measures were highly correlated.

3. Analyses of juvenile overwinter survival were conducted using mixed-model logistic regression (GLIMMIX macro and PROC MIXED in SAS). I used survival from age 30 days until the next nesting season as the binomial response variable here. Analyses of juvenile survival were conducted in the same general way as analyses of adult survival, as described in note 2. As with analysis of data on adult survival, analyses were conducted in three stages in order to minimize the number of models within a given set. Models in the first set (see the chapter's digital appendix, available from author, for details) examined whether juvenile and adult population densities affected survival probabilities. Nesting date was forced into all models based on prior knowledge of the system (Hochachka 1990), as were the effects of three supplemental feeding experiments that occurred in 1979, 1985, and 1988. There were three well-supported models in this first set.

To each of these three models from the first step, I added effects of mothers' and/or fathers' ages on the survival of juveniles. Each parent's age was described with either a linear effect or linear plus quadratic effects. The result was a model set containing 21 different models (see the digital appendix) for the second step in the analysis.

There were five well-supported models in this second step, three of which contained effects of fathers' age on juvenile survival. These five models were then used to assess whether full siblings had similar probabilities of survival. The previous two sets of models contained two random effects, year and family ID (a unique identifier for each set of full siblings), and model fitting was conducted using maximum likelihood analyses. In the final set of analyses, we took each of the five combinations of fixed effects identified above, and examined the support for three different sets of random effects: year only, year plus family ID, or year plus annual family ID (each set of full siblings born within a single year possessed a unique identifier). Within each combination of fixed effects, the relative support for each of the three sets of random effects was compared based on Akaike weights from restricted maximum-likelihood analysis. In all cases, the best supported random effects were the combination of year plus year-specific family ID. The final five sets of fixed effects, with the best-supported pair of random effects, were then used for graphing, with the plotted predicted values being calculated by model averaging across the five models.

4. I looked for spatial-autocorrelation in the probability of nest failure in two stages. First, I created a statistical model that predicted probability of nest failure but that did not account for the possibility of nearby nests having the

same fate. Second, I examined the residuals from this initial model to see if residuals were more similar at shorter distances between nest locations. As with the other analyses in this chapter, the dependent variable, nest failure, was binomially distributed. I thus used mixed model logistic regression using the GLIMMIX macro in SAS. Female age, date of nest initiation, clutch size, and the stage at which a nest was found were all assumed to affect probability of nest failure and were entered into the initial statistical model.

To explain additional year-to-year variation in the probability of nest failure, I either treated year as a categorical variable or used the number of breeding females (i.e., population density) and presence of cowbirds in the island (a binomial variable) as the underlying causes of most year-to-year variation. Based on Akaike weights, there was no support for using year as a categorical variable. That is, the effects of population density and the presence of cowbirds could explain most year-to-year variation in annual rates of nest failure. In this comparison of fixed effects, we used female ID as a random variable in order to account for within-female consistency in probability of nest failure. However, the likelihood that a female's nests will fail may only be consistent within a single year and not across years. Thus, I contrasted the support for female ID and year-specific female ID as random variables. Essentially all support was for using female ID, indicating that the probability that a female's nests would fail was consistent both within and among years.

Residuals from this final model (on the logit scale) were then compiled, and the distances between all possible pairs of nests were calculated. As an indicator of correlation in residuals, the absolute value of the difference between paired residuals was calculated. Smaller numbers indicate higher similarity. I classified pairs of nests as being from the same year, or 1, 2, 3, or >3 years apart. This was done in case "poor" areas on Mandarte shifted through time.

10 Population Viability in the Presence and Absence of Cowbirds, Catastrophic Mortality, and Immigration

Peter Arcese and Amy B. Marr

Catastrophic mortality and density dependent juvenile mortality outside the breeding period limit the population size of song sparrows on Mandarte Island. The presence of brood parasitic brown-headed cowbirds reduces reproductive rate in the population. These factors were explored individually in preceding chapters. Here, we explore how catastrophes, juve-

*nile mortality, cowbirds, and a fourth factor, immigration,
interact to influence the long-term size and viability of the
Mandarte Island song sparrow population.*

Isolation, catastrophic mortality, small average population size, and exposure to nonnative enemies all influence the dynamics of Mandarte
Island song sparrows (see chapters 3–5). Several populations of song
sparrows of concern to conservation also face similar threats (Chan and
Arcese 2002), as do other small and isolated populations of vertebrates.
In this chapter we describe a simple model to rank the potential influence of some of these factors on the expected size and viability of the
Mandarte Island song sparrow population. Because catastrophic mortality events affect all ages of song sparrows on Mandarte Island, we expected that the frequency of catastrophes would have an overriding influence on population viability. We also expected that immigration to
the island might occasionally rescue the population from extinction. Finally, because cowbirds reduced the reproductive rate of song sparrows
on Mandarte Island, we expected their presence to reduce the average
population size and increase the risk of extinction. Below, we begin by
providing some background to demographic analysis and the assessment
of potential threats in the context of population viability analysis (PVA).

Whenever we try to conserve populations of plants or animals, questions naturally arise about the hierarchy of threats facing those populations, the alternative approaches to ameliorate those threats, and the
likelihood that the amelioration of key threats will result in recovery
(Beissinger 2002; Boyce 2002). Answering questions such as these requires that we understand the mechanisms of population growth and
stability and their influence on population persistence. So far, however,
few populations worldwide have been studied in sufficient detail to answer such questions with precision (Ludwig 1999; Beissinger 2002).

One way to improve understanding is to use the results of detailed
demographic studies to simulate the response of a population to hy-
pothetical threats using PVA (Boyce 1992, 2002; Beissinger and Westphal 1998; Beissinger and McCullough 2002; Reed et al. 2002). As
summarized in the reviews cited, PVA has gained general acceptance
as a management tool because of its advantages over earlier and less
rigorous approaches. The main use of population viability models lies
in the fact that constructing them forces biologists and managers to be
explicit about potential threats, the quality of the data in hand, and the
key assumptions underlying management (Walters 1986; Hilborn and

Mangel 1997). At a minimum, these steps should help recovery teams to identify what research is required before reliable recovery plans can be formulated.

The reviews cited above, however, also identify difficulties with PVA. For rare, cryptic, and long-lived species, data and life history information are often sparse or missing entirely. In a few cases, hypothesized threats to populations have been studied experimentally to identify limiting factors and management responses (Caughley and Gunn 1996; Smith et al. 2002), but sociopolitical obstacles often prevent experimental approaches in practice (Walters 1997; Ludwig and Walters 2002). Thus, even in carefully studied populations, uncertainties about life history and demography can make it hard to predict a population's response to management, and harder to estimate its likelihood of extinction in the absence of management (e.g., Hilborn and Mangel 1997; Sæther et al. 2000a). Thus, there is often a gap between the promise of PVA and how well the predictions derived from it perform in practice. Because of these difficulties, new approaches are being employed to improve the predictions of PVA.

One such approach is to employ the results from studies of "model species" to reduce uncertainty about demographic processes and develop ranked lists of threats to comparable "species at risk." Studies of model species can help us to refine estimates of uncertain demographic rates (e.g., Ralls et al. 2002; Lande et al. 2003) and assess the relative strength of various intrinsic and extrinsic effects on populations (Keller and Waller 2002; Arcese 2003). Demographic rates and the mechanisms underlying population growth and stability are often well described for species that have been the subject of detailed, longitudinal study. Examples include some species at risk (e.g., Woolfenden and Fitzpatrick 1989; Reid et al. 2003b). Studies of the Mandarte Island song sparrow population might facilitate the management of threatened song sparrow populations isolated in remnant patches of historically extensive habitats (Chan and Arcese 2002). Our results can also be used to speculate about the influence of particular extrinsic threats within the general framework of PVA.

Three extrinsic threats of particular interest to managers of rare species are abundant predators or parasites, the curtailment of natural immigration, and the frequency and severity of natural or human-caused catastrophes (Boyce 1992; Beissinger and Westphal 1998; Beissinger and McCullough 2002). All three factors have the potential to limit the size of the resident sparrow population on Mandarte Island (Arcese et al. 1992, 1996; see chapters 3–5). We explored the influence of these factors on population size and viability by using a simple population model and alternative assumptions about the presence or absence of brown-headed cowbirds, natural immigration, and catastrophic episodes of mortality. We also explored the effects of including stochastic error in reproduction and juvenile survival on population performance.

10.1. Modeling Approach and Predictions

We set out to rank the relative influence of brown-headed cowbirds and variation in overwinter survival and immigration rate on the persistence and size of the Mandarte Island song sparrow population. To do so, we built a simulation model of the population based on the number of breeding females alive in spring, their reproductive and survival rates, and the survival to recruitment and breeding of independent young (figure 10.1; Arcese et al. 1992; see chapters 3–5). Our previous work had shown two results. First, reproductive rate is affected markedly by the presence and intensity of brood parasitism by brown-headed cowbirds (Arcese et al. 1992, 1996; Arcese and Smith 1999; see chapter 5). Second, population size is closely correlated with the local survival of adults and juveniles, especially as a consequence of episodes of severe mortality over winter (Arcese et al. 1992; see chapters 3, 4). Thus, we expected the presence of cowbirds to reduce mean population size and increase the likelihood of extinction via demographic stochasticity.

Although immigrants to Mandarte are uncommon, they can comprise a substantial fraction of the population at small sizes (Marr et al. 2002; see chapters 4, 8). Thus, it is reasonable to expect that immigrants may facilitate the demographic rescue of small populations by augmenting their size (Gilpin and Soulé 1986). Finally, because the introduction of stochastic error in demography destabilizes populations and increases the likelihood of extinction in simulation models (Lande et al. 2003), we also explored the consequences for population persistence of adding stochastic error in reproduction and juvenile survival rates, relative to the influence of cowbirds, demographic rescue, and catastrophic mortality.

The Population Model

We employed a balance equation model (Walters 1986) to estimate the fraction of populations that became extinct over 500 trials (<1 fe-

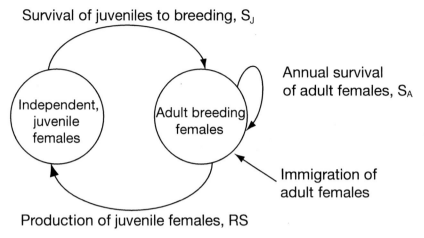

Survival of juveniles to breeding, S_J

Independent, juvenile females

Adult breeding females

Annual survival of adult females, S_A

Immigration of adult females

Production of juvenile females, RS

Figure 10.1. A life cycle diagram used to model the Mandarte Island song sparrow population. Model steps are indicated by arrows and begin with a spring population census. Immigrants then augment spring population size, followed by the annual production of independent young. Overwinter survival rates of adult and juvenile females are then applied to generate population size in the subsequent year (details in text). The model written as a balance equation is $N_{t+1} = (N_t \times S_A) + (RS \times S_J)/2 +$ female immigrants. The model was built to estimate the relative influence of cowbirds, immigration, and natural catastrophe on the persistence of the Mandarte population in future.

male in spring) and the expected distribution of population sizes 100 years into the future. Key time steps in the model included the annual production of independent young, the recruitment of independent young to the breeding population in the following year, the survival of breeders to the next year, and the annual number of female immigrants that arrived and bred (figure 10.1). These time steps approximate life history stages and events identified elsewhere as influencing demography (see chapters 3–5). The model followed a prebreeding format, wherein population size was evaluated annually in spring, after winter mortality had occurred, but prior to the production of young (figure 10.1). Adult mortality in summer was a small fraction of annual mortality and not modeled separately (Arcese et al. 1992). Initial population size was taken as 45 breeding females. We modeled only females for simplicity and because other work suggests that male numbers and adult sex ratio have minor effect on the dynamics of the population (Arcese 1989a, 1989c; Arcese et al. 1992; see chapters 3, 4, 6). We now describe how we estimated the parameters in figure 10.1.

Annual Survival of Adult Females (S_A)

We modeled the annual survival of adult females as a pseudorandom variable drawn with replacement from the observed distribution of annual survival rates in the population (figure 10.2; see chapter 3). Our method restricted the range and frequency of modeled survival rates to those observed in data, and it assumes that adult mortality rates were independent with respect to year (see section 10.3). Other methods of estimating survival from these data, such as drawing from one or more probability distributions fitted to them (e.g., Ludwig 1999), gave similar results to those shown here. It is possible that predictive relationships of adult survival based on factors such as population density, age structure, or inbreeding will be developed in the future (e.g., see predictions for S_J below), but they were not employed here because, with the exception of inbreeding (see chapter 7), they were not supported by data (chapter 9).

Annual Survival of Independent Female Young (S_J)

The fraction of independent females that survived to breed in the year after hatching declined as the number of breeding females in spring increased (Arcese et al. 1992; see chapters 3, 4). However, juvenile females also suffered the same density-independent episodes of catastrophic mortality that were experienced by adults in the same year (Arcese et al. 1992; see chapter 3). To preserve this correlation of adult and juvenile female survival in our model, we estimated S_J from the regression of observed juvenile survival on the number of adult females in spring and the observed survival of those adult females to the next year:

$$S_J = 0.24 - 0.004 \times \text{adult females} + 0.43 \times S_A + \text{error} \quad (1)$$

This regression accounted for 58% of the variance in juvenile survival.[1]

We investigated the additional effect of error in estimates of S_J on population performance by drawing annual estimates either deterministically from equation 1, or using deterministic values plus or minus random values drawn from a normal distribution with mean = S_J and SD = $0.1 \times S_J$. For example, for a deterministic estimate of $S_J = 0.10$ (equation 1), our assumption of 10% error implies that, in about 67% of years, the model will return values of $S_J = 0.09–0.11$. In about 1% of years, values as high as 0.13 and as low as 0.07 will be observed

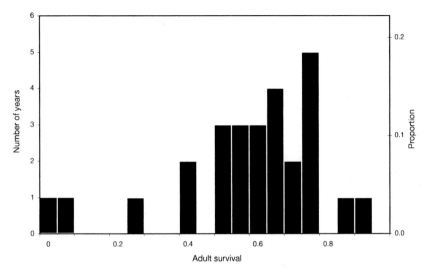

Figure 10.2. Frequency histogram of values of adult survival for female song sparrows observed on Mandarte Island over 27 years (see also figure 3.8; range = 0.04–0.91, median = 0.64, mean = 0.60). Two years of "catastrophic" decline occurred in 1979–1980 ($S_A = 0.07$) and 1988–1989 ($S_A = 0.04$).

($S_J \pm 3$ SD). We also imposed lower and upper bounds of 0.04 and 1.0 on S_J to maintain values within the range of those possible and observed historically.

The addition of stochastic error to population models can simulate the influence of undetected and independent random variables affecting survival, measurement error, or both (Lande et al. 2003). Increasing the variance in survival rate is expected to raise extinction rates in the wild, whereas measurement error that overestimates variation in vital rates may cause the overestimation of extinction risk (e.g., Gilpin and Soulé 1986; White et al. 2002).

Catastrophic Population Decline

Episodes of severe overwinter mortality in young and adult song sparrows have an overriding influence on population size on Mandarte Island (Arcese et al. 1992). In 1979–1980 and 1989–1990, 93% and 96% of adult females disappeared from the island, respectively (figure 10.2; see chapters 3, 4). Although "catastrophic" episodes of mortality occurred just twice in our 28-year study, this implies a 7% chance of observing a catastrophe in any given year in future and a 0.5% (0.07 squared) chance of catastrophes in 2 consecutive years. Catastrophic mortality is of obvious relevance in PVA because it has a profound influence on the likelihood of extinction (Ludwig 1999). We therefore quantified its effect in our analyses here by removing from our resampling procedure for S_A the most severe record of mortality in 1988–1989, when adult females suffered 96% mortality. By doing so, we reduced the expected annual occurrence of catastrophes from 7% to 3.7% and the expected occurrence of two consecutive catastrophes to 0.1%.

Reproductive Output in the Presence and Absence of Cowbirds

We estimated the total number of independent young raised annually (reproductive output; RO) from data collected in years with brown-headed cowbirds absent or present on the island (e.g., figure 10.3a; see chapter 5). In the absence of cowbirds RO increased linearly with the number of females breeding, with no evidence of density dependence (figure 10.3a see chapter 5). In contrast, reproduction declined strongly above median density in the presence of cowbirds and was predicted well by quadratic regression (figure 10.3a).

A tendency for cowbirds to visit Mandarte more often at high density since 1974 has prevented us from studying the population at high density in their absence (see chapter 5). However, Tompa (1963) reported four observations of reproduction for Mandarte Island song sparrows from 1960–1963, before cowbirds colonized the island. Tompa's data straddled the regression line fitted to our own data in the absence of cowbirds (figure 10.3a). We therefore used data collected after 1974 to estimate the overall impact of cowbirds on the population via their effects on annual reproduction (figure 10.3b).

We estimated RO in the presence and absence of cowbirds by fitting a quadratic regression to each data set while forcing the intercept through zero and including females and females^2 as predictors. The initial fits differed mainly in the size of the quadratic regression coefficient b_2 (females squared) and to a small degree in size of the linear regression coefficient b_1 (figure 10.3a). To complete our estimation procedure, we accepted the average of b_1 values (weighted by sample size), and then values of b_2 fitted by maximum likelihood and least squares (Hilborn and Mangel 1997). This allowed us to fit each data set using a quadratic relationship that differed only in the magnitude of density

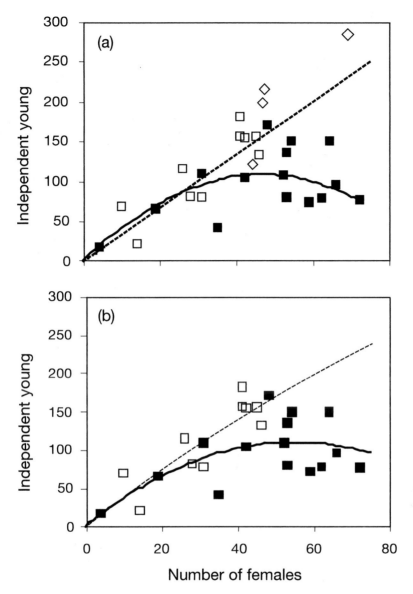

Figure 10.3. Observed numbers of independent song sparrow young produced by different numbers of breeding females in the presence (solid squares) versus absence (open squares) of brown-headed cowbirds. (a) Data, including four observations made from 1960 through 1963 by Tompa (1963; open diamonds). Lines indicate the linear fit to the data in the presence and absence of cowbirds after 1974 (see note 2). (b) The expected number of independent young produced by different numbers of breeding females in the presence (solid squares) versus absence (open squares) of brown-headed cowbirds. The fitted relationships were forced through the origin and differ only in the magnitude of the quadratic regression coefficient (female squared; see text).

dependence (b_2: females; figure 10.3b).[2] Thus, in the absence of cowbirds reproduction was estimated as

$$RO = 3.97 \times females-0.011 \times (females)^2 \qquad (2)$$

In the presence of cowbirds reproduction was estimated from Equation 2 after substituting the value -0.036 for the coefficient of the quadratic term, females.[2]

We modeled stochastic error in annual estimates of annual reproduction following the method described above for S_J. We did not model cowbirds effects on S_A or S_J because none have been detected (see chapter 5).

Natural Immigration and Demographic Rescue

We used data on the number of female immigrants that bred on Mandarte to estimate the effect of demographic rescue on the persistence of the population relative to other potential threats. To do so, we followed the method described above for S_A. Over 28 years, we observed no female immigrants in 12 years, 1 in 10 years, 2 in 5 years, and 3 in 1 year. Thus, the probabilities of drawing 0, 1, 2, or 3 immigrants in a given year were approximately 0.43, 0.36, 0.18, and 0.04, respectively. The probability of drawing at least one immigrant in a given year was 0.57.

Model Scenarios

We explored four scenarios, each in the presence or absence of cowbirds, to rank their effects on population performance. First, an "island case" considered the Mandarte Island population in the presence of immigration, catastrophic mortality, and stochastic error in S_J and reproduction in the absence and presence of cowbirds. Second, a "no error" case mimicked the island case, excluding stochastic error in S_J and reproduction. Third, a "no 1989 crash" case included immigration, stochastic error, and the influence of cowbirds, but it omitted from our sample distribution for S_A the severest crash during our study (1989; figure 10.2). Fourth, a "no rescue" case mimicked our island case in the absence of immigration to the island.

10.2. Results

The presence of brown-headed cowbirds increased extinction risk markedly in comparison to cases modeled in their absence (figure 10.4a). The expected mean of population sizes was also reduced

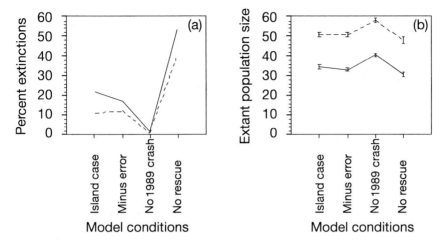

Figure 10.4. Percentage of populations becoming extirpated (a) and mean (±SD) size of extant populations (b). Estimates were based on the results of 500, 100-year model runs. Model conditions are described in text. Solid lines depict results based on the expected number of independent young produced in the presence of brown-headed cowbirds, and dashed lines depict results expected in the absence of cowbirds (see figure 10.3b and text for a description of estimation procedure).

39–55% in the presence of cowbirds (figure 10.4b). However, catastrophic mortality and immigration each had a greater influence on model outcomes. In the presence of immigration, catastrophes, cowbirds and error in S_A and RO ("island case"), 11% of simulations ended in extinction, compared to 22% in the presence of cowbirds (figure 10.4b). In contrast, removing stochastic error in estimates of RO and S_J had little effect on either the size of extant populations (figure 10.4a) or their expected rate of extinction (12% in the absence and 17% in the presence of cowbirds; figure 10.4b).

In contrast, removing the 1989 record of catastrophic mortality from our sampled values of S_A reduced the extinction rates dramatically, to 0.6% and 1% in the absence and presence of cowbirds, respectively. Reducing the chance of catastrophic mortality also increased expected population size with cowbirds absent and present (Mean ± SD = 58 ± 1 and 41 ± 1, respectively; figure 10.4b) because populations were only half as likely to suffer catastrophic reductions in size.

Preventing demographic rescue via immigration also caused extinction rate to increase dramatically to 39% and 59% of runs with cowbirds absent and present, respectively (figure 10.4a). In the absence of immigration, mean population size was also reduced versus the comparable island case with cowbirds absent (48 ± 2 vs. 51 ± 1, immigrants absent vs. present, respectively) or present (31 ± 1 vs. 33 ± 1; figure 10.4b).

Overall, our results provide a clear picture of the expected influence of extrinsic effects on the performance of the Mandarte Island song sparrow population. The magnitudes of effects, however, depended somewhat on the performance measure used. In the case of extinction risk, the rank order of influence from most to least severe was catastrophic mortality, immigration, cowbird impacts, and stochastic error in S_J and RO. In contrast, cowbirds had the largest negative effect on population size and increased the expected risk of extinction as a consequence (figure 10.4a,b). The remaining extrinsic effects on N in order of influence were catastrophic mortality, immigration, and stochastic error in S_J and RO.

10.3. Discussion

Catastrophes and Extinction

The frequency of catastrophic mortality played a key role in the extinction dynamics of our model population. Extinction coincided with catastrophes, particularly when drawn in consecutive years. Thus, removing the more severe of two catastrophes from our 28-year sample distribution reduced the expected extinction rate to $\leq 1\%$ (figure 10.4). Clearly, our approach to modeling S_A by sampling from the observed distribution of values also introduced substantial demographic stochasticity into our model. Stochastic variation in adult mortality and its correlated effect on S_J swamped the effects adding stochastic error to RO and S_J (figure 10.4). Overall, these results support those of Ludwig (1999), who showed that uncertainty in the frequency of catastrophic mortality had an overwhelming influence on the expected rate of extinction in this population.

Pimm and Bass (2002) also report that small differences in the assumed frequency of wildfires had large effects on the expected population size of Cape Sable seaside sparrows. Unfortunately, the history of study for most populations is short, and data on the frequency of catastrophe are still rare. Moreover, it is reasonable to expect the frequency of catastrophe to vary by location or, perhaps, population state (Arcese 2003).

Biologists and managers generally respond to the possibility of catastrophe by adopting a precautionary approach to the estimation of minimum viable populations and population persistence (Beissinger 2002). Ralls et al. (2002) also recommend that the consequences of poorly described parameters in PVA be examined explicitly but also suggest that default values might be used to harmonize the results of PVA for little known species. Although it is possible that enough data will eventually accumulate to estimate the frequency of catastrophic mortality accurately in some populations, extrapolation may also become more complicated as links between individual phenotype and population demography are discovered. In Soay sheep, for example, weather and density each affect populations directly but also indirectly via their interaction with population age structure, sex ratio, and individual fitness as a consequence of additive genetic variation in body size, parasite resistance, and feeding apparatus (Coltman et al. 1999; Smith et al. 1999; Milner-Gulland et al. 2000). These interactions contribute to predictable dynamics that will affect models designed for PVA.

In our current model, the frequency of catastrophe was made independent of year and population state. However, it is plausible that deterministic factors also affect the probability of catastrophe in the Mandarte Island song sparrows. In particular, individual-by-environment interactions, such as those arising from correlations between the mean state of individuals and size of populations, may influence the frequency and severity of catastrophic decline. As an example, inbreeding depression can be expressed as a nonlinear, additive mortality during periods of severe environmental stress (Keller et al. 2002; Keller and Waller 2002). On Mandarte Island, episodes of severe overwinter mortality have occurred at roughly 10-year intervals since 1979 (see chapter 3). It is therefore interesting to speculate that periodic mortality might arise as a deterministic consequence of interactions between an intrinsic population trait, such as its size or level of inbreeding, and extrinsic limiting factors such as severe weather, disease, or predation (Arcese 2003). Keller (1998; see chapter 7) described marked selection against inbred sparrows during the 1989 crash and a rapid rise in the mean level of inbreeding subsequently due to the small average population size and low rate of immigration. If inbreeding depression is expressed as a threshold effect in relation to environmental stress, the probability of catastrophe will depend on population history and yield dynamics reminiscent of those observed on Mandarte (see chapter 4; S. E. Runyan and P. Arcese, unpublished results).

In practice, the successful management of populations subject to regular catastrophe will rely on the maintenance of multiple populations separated in space and on ensuring that dispersal takes place among these populations (reviewed in Hanski 2002; Harrison and Ray 2002). As predictors of catastrophic loss are identified, managers may be able to take measures to avoid them. In patchily distributed populations like those on Mandarte Island, minimizing catastrophic losses that occur as a consequence of a deterministic links between degree of inbreeding and the susceptibility of individuals to extrinsic stressors might be managed by maintaining natural dispersal or translocation (e.g., Westemeier et al. 1998; see chapter 7).

Immigration and Demographic Rescue

After catastrophic mortality, the presence or absence of immigration to the population had the largest effect on extinction risk but a smaller effect on expected population size (figure 10.4a,b). These results suggest that small numbers of immigrants may enhance substantially the persistence of populations subject to catastrophe. It is also interesting that we found such a pronounced effect of immigration solely via demographic rescue. Immigrants increased the size of very small populations substantially, despite a relatively small effect on mean population size (figure 10.4b). Immigrants also contributed to the production of new recruits and accelerated population growth. Because immigrants were counted in the spring census, they also contributed directly to the "recolonization" of the model population in years wherein no resident adult or young survived. In nature, the frequency of recolonization to patch occupancy can only be estimated precisely when residents and immigrants are uniquely marked. Because complete enumeration is rarely accomplished, studies of demographic rescue in nature will remain challenging. Our results suggest that immigration provides a buffer against the effects of stochastic variation in adult and juvenile mortality on the probability of extinction.

In addition to the direct effects of immigration on population persistence, much emphasis has also been placed on the influence of immigration via its effects on inbreeding and genetic diversity (reviewed in Hedrick and Kalinowski 2000; Allendorf and Ryman 2002). Immigrants to the Mandarte song sparrow population caused the rapid recolonization of extinct alleles following catastrophic decline (Keller et al. 2001; see chapters 7, 8) and by contributing offspring with superior fitness to the breeding population (Marr et al. 2002). Inbreeding also has affects the fitness of individual sparrows, which can be expected to act cumulatively to reduce population growth as the level of inbreeding rises (Keller 1998; Marr et al. 2002; see chapters 7, 8). As a consequence, models of the Mandarte Island population that incorporate inbreeding effects should show more pronounced effects of immigration on persistence than reported here, due to the positive influence of genetic rescue from inbreeding depression on survival and reproduction.

Brown-headed Cowbirds and Population Performance

Our results also suggest that the arrival of cowbirds in the Georgia Basin of southern British Columbia and on Mandarte Island has had deleterious effects on the demography of song sparrow populations. Brown-headed cowbirds invaded the Georgia Basin in the 1960s and first bred on Mandarte after 1963. Data collected on Mandarte by Tompa (1963), before cowbirds arrived, support our comparative results which show that density-dependent declines in reproduction arise mainly as a consequence of high nest failure in the presence cowbirds (figure 10.3; Arcese et al. 1992, 1996; Arcese and Smith 1999). Smith et al. (2002) supported these results by showing that song sparrows in rural landscapes about 50 km northeast of Mandarte Island were unable to reproduce at levels sufficient to maintain population growth, but experienced about a 2-fold increase in reproduction and expected growth when parasitism was reduced by about half. Overall, these results show that cowbirds can have marked deleterious effects on individual and population fitness in the song sparrow. They also suggest that estimating reliably the influence of cowbirds on their hosts requires that biologists employ comparative or experimental approaches to contrast population performance in the presence and absence of cowbirds (Arcese et al. 1996; Arcese and Smith 1999).

Cowbirds imposed a density-dependent cap on the rate of reproduction in our model, whereas reproduction declined only slightly as density increased in the absence of cowbirds (figure 10.3; see chapter 6). Capping reproduction reduced the expected size of the model population and increased the rate of extinction (figure 10.4a,b). It is axiomatic that population size will be related negatively to extinction risk in populations subject to catastrophic mortality (e.g., Gilpin and Soulé 1986; Lande et al. 2003).

Brown-headed cowbirds visited Mandarte Island less often during the latter third of our study (see chapter 6), perhaps as a consequence of habitat change (Saunders et al. 2003). Thus, a simple interpretation of our model implies that we might also expect to have observed an increase in population size during that time. However, because variation in survival is the dominant factor regulating population size (Arcese et al. 1992; see chapter 4), and long-term variation in climate drives annual variation in RO (Wilson and Arcese 2003), an accurate partitioning of the components of variation in population size will require additional study. Nevertheless, our current results show that the size and growth of the Mandarte Island song sparrow population will each be reduced on average in the presence of cowbirds (figure 10.3a,b). Because cowbirds reduce reproduction in song sparrow populations, they may also reduce immigration in regional meta-populations.

Researchers have traditionally estimated the cumulative impact of cowbirds on host and population fitness by comparing the number of young fledged from parasitized and nonparasitized nests or females within populations (e.g., Pease and Grzybowski 1995). This approach assumes that cowbirds have their main influence on hosts via egg removal, nest desertion, and the usurpation of parental care (Arcese et al. 1996). By failing to consider the effect of cowbirds on nest depredation or other modes of failure not often attributed to brood parasites, researchers risk missing the largest deleterious impact of cowbirds on host reproduction. On Mandarte Island, Smith (1981a) estimated cowbird impacts on host fitness by comparing parasitized and nonparasitized hosts within years, and he concluded as a consequence that cowbirds had little influence on host fitness. Comparative and experimental work has since demonstrated large cumulative effects of cowbirds on host and population fitness (Arcese et al. 1992, 1996; Arcese and Smith 1999; Smith et al. 2002; see chapter 5). Overall, our results raise questions about estimates of cowbird impact based on traditional methods, and they suggest that cowbird impacts are often underestimated.

Our results also suggest, however, that the management response to cowbirds should depend on their influence on host demography, relative to other factors that might be managed (Arcese 2003). Our current results show that cowbird impacts on our modeled song sparrow population were substantial, but also that reducing the risk of catastrophic loss and isolation from immigration may have larger beneficial effects on population persistence than removing cowbirds.

10.4. Conclusions

Populations are affected by a hierarchy of extrinsic and intrinsic factors and their successful management should be helped if we understand the mechanisms of population growth and stability and their influence on persistence. We developed a simple PVA for a single well-studied population that might be used help prioritize management efforts in ecologically similar species that cannot be studied in detail. Not surprisingly, the frequency of catastrophic mortality had an overwhelming effect on the frequency of extinction in our model. Immigration, via demographic rescue, also had a positive effect on the recovery from small population size and rate of extinction. Brown-headed cowbirds reduced expected population size and raised the probability of extinction. If cowbirds also reduce the rate of dispersal between populations by reducing reproductive rates within them, cowbirds will have more pronounced effects on populations than estimated here, because their presence will also reduce the positive genetic and demographic effects of immigration. We also suggest that interactions among small population size, inbreeding, and inbreeding by environment interactions affecting fitness (see chapter 7, 8), which we did not model here, are likely to affect the size and stability of small populations.

Notes

1. Best-fit regression to estimate the local survival of juvenile song sparrows from independence to April of the year following hatch ("recruitment"): Coefficients are given with their associated standard errors: $S_J = 0.237$ (0.092) $- 0.00429$ (0.001) $\times$ females $+ 0.432$ (0.11) $\times$ survival of adult females (adjusted $r^2 = 0.58$, $F_{2,25} = 18.06$, $p < 0.001$).

2. Best-fit regressions used to estimate the expected number of independent young raised on Mandarte Island in the presence and absence of brown-headed cowbirds: Regressions were forced through the origin. Coefficients are given with their associated standard errors. Cowbirds present: RO = 4.46 (0.85) $\times$ females $- 0.046$ (0.015) $\times$ females^2 ($F_{2,15} = 66.8$, $p < 0.001$); cowbirds absent: RO = 3.35 (0.27) $\times$ females ($F_{1,12} = 153.8$, $p < 0.001$).

11 Genetic and Demographic Risks to Small Populations Revisited

James N. M. Smith, Lukas F. Keller,
and Jane M. Reid

A realistic integration of demography and population genetics, applicable to species in natural environments, is a formidable task that has enticed but largely eluded ecologists and evolutionary biologists. (Russell Lande 1988a, p. 1459)

Our principal theme in this book has been to describe how small size affects the demographic performance and genetic composition of small populations. We began with a brief history of the genetic and demographic risks that are likely to affect small populations. We considered these ideas through a detailed example, the song sparrows of Mandarte

Island, and by reviewing recent advances in broader knowledge. We hope we have achieved a detailed description of the genetics and demography of this focal population and that we have gained insights relevant to the biology and management of small populations in general.

Since Lande (1988a) pointed out the difficulty of developing a satisfactory theory linking demography and population genetics, our knowledge has grown rapidly. The theory underpinning Caughley's (1994) small population paradigm has been developed (e.g., Lande et al. 2003; Engen et al. 2003) and applied to small populations in the wild (e.g., Nieminen et al. 2001; Clutton-Brock and Pemberton 2004). Our knowledge of the effects of inbreeding in the wild has grown enormously (Keller and Waller 2002), and advances have also been made in other branches of conservation genetics (Frankham et al. 2002). While genetic rescue of small and struggling populations has been achieved (Ebert et al. 2002; Vilà et al. 2003), we also know that such attempts can fail disastrously (Hughes et al. 2003). Finally, we are improving our understanding of how genetics and ecology interact to affect populations (Saccheri et al. 1998; Keller et al. 2002).

In this concluding chapter, we summarize what we have learned about the biology of the population of song sparrows on Mandarte Island and consider the applicability of our findings to other small populations. We reconsider the demographic risks facing small populations and assess the genetic risks. In this way, we return to the question first raised by Lande (1988a): Are genetic threats to small populations of secondary concern compared to demographic or environmental risks? We conclude with some advice for managers of small populations and questions for discussion by student readers.

11.1. Demographic Risks to Small Populations

Environmental Variability, Density Dependence, and Population Recovery

Small populations experience both demographic and environmental stochasticity (Shaffer 1981). One way to assess the relative impact of these processes is to calculate the population size at which the effects of environmental stochasticity on population growth rate are 10-fold higher than those of demographic stochasticity (p. 9 in Lande et al. 2003). Above that population size, the effects of demographic stochasticity on population trajectories can be neglected. Using information in Sæther et al. (2000a) we calculated this threshold size to be 16 females for

Mandarte song sparrows. This number was exceeded in all but 6 of our 28 study years. Moreover, even when the population reached its lowest level (four females in 1989), environmental stochasticity still contributed 2.5 times as much to the variance in growth rate as did demographic stochasticity. Thus, as expected from theory (Shaffer 1981), demographic stochasticity influenced the song sparrow population growth rate on Mandarte in only a few years and was otherwise negligible. Environmental stochasticity, however, had a major influence.

The impact of environmental variability is reflected in the fact that song sparrows on Mandarte are one of the least stable bird populations studied to date (table 1.1 in Lande et al. 2003). They rebounded twice from catastrophes that reduced the numbers of females to 10–20% of the median population size (see figure 4.1). The population was able to rebound from these bottlenecks mainly because juvenile survival increased at low densities (figures 4.4, 9.3). Immigrants to Mandarte also helped the population to recover (figures 4.6, 10.4). The substantially larger Capricorn silvereye population on Heron Island, Australia, also rebounded rapidly after population bottlenecks, at least partly due to improved juvenile survival (McCallum et al. 2000).

In our study we found no evidence of positive density dependence (e.g. Allee effects) in reproductive success or survival over the first 23 years of our study. Since 1998, however, the population has exhibited variable reproductive success and low juvenile survival at below-average densities. These observations could indicate Allee effects. However, we can only speculate whether this recent failure to rebound from low densities may reflect habitat change (chapter 2) and/or temporal changes in the effects of inbreeding (figure 7.4) rather than an intrinsic consequence of small population size per se.

Simple negative density dependence is not the only possible response to a sudden drop in population size. Island-dwelling Soay sheep exhibit more complex responses to population crashes than do either song sparrows or silvereyes, since survival and fecundity depend on fluctuating weather, age structure, and density (Coulson et al. 2001). Other populations have declined to extinction after becoming small (see figure 2.3 in Lande et al. 2003), as envisioned in the extinction vortex. In these last cases, standard inverse density dependence clearly did not rescue the population.

We currently lack a good understanding of why populations of some species rebound quickly when reduced to a small size while others do not. Potential explanations include: intrinsic features of a species' life history, interactions between life history variables and the environment

(Clutton-Brock and Coulson 2002), and genetic changes. The recent failure of Mandarte song sparrows to rebound from low densities occurred during a time of reduced immigration and hence lower genetic exchange compared to 1981–1992 (figure 4.6).

Unstable Long-term Dynamics

Aspects of the population dynamics of Mandarte's song sparrows, including the form of density dependence, changed during our study (chapters 4–6). First, cowbirds visited the island less often after 1989, and density-dependent reproductive success was no longer evident in years when cowbirds were absent and nest failures became less frequent (chapter 5). Second, poor survival of juvenile females late in our study (1992–2001) biased the adult sex ratio (figure 4.1), reduced effective population size, and influenced the mating system (chapter 6). The low local survival rates of juvenile females might have been a consequence of their low dominance status in late summer, autumn, and winter (chapters 6, 9). These shifts only became evident after many years of study (compare this summary to results of Arcese et al. 1992).

Temporal instability in numbers exists in many other animal populations. Florida scrub-jays (Woolfenden and Fitzpatrick 1991), European dippers (Sæther et al. 2000b), kangaroos and caribou (Caughley and Gunn 1993), Galápagos finches (Grant et al. 2000), and Soay sheep (Coulson et al. 2001) all fluctuate markedly. Invertebrate (Andrewartha and Birch 1954) and amphibian (Meyer et al. 1998) populations can fluctuate even more markedly and erratically. Spatial patterns in dynamics also vary widely within species (Krebs 2002). These findings should worry conservation biologists since they recall Ludwig's (1999) warning about the dangers of basing predictions of future population performance on short runs of data.

Effects of an Invader

When mammalian predators are introduced to islands, they can affect native species severely (Blackburn et al. 2004). Invading birds are seldom seen as a threat, although the Indian mynah, a human camp follower, may threaten some island endemics (Komdeur 1996). However, one avian invader, the brown-headed cowbird, has received detailed attention (Smith et al. 2000). The cowbird colonized much of North America over the past 200 years (chapter 5). It reached southern Vancouver Island in about 1955 (Rothstein 1994). The cowbird was almost an annual visitor to Mandarte Island early in our study, but then it visited less frequently (chapter 5).

We initially concluded that the cowbird had a negligible effect on the sparrow population (Smith 1981a). However, after further study, we revised this opinion (Arcese and Smith 1999). We now believe that cowbirds affected the population dynamics of song sparrows strongly. They were largely responsible for driving the density-dependent reproductive success we observed in song sparrows through their effects on fecundity and nest failure (chapters 4, 5); they also depressed population size and increased the probability of population extirpation (chapter 10).

Our results here generalize well to other populations of song sparrows (Arcese and Smith 1999; Zanette et al. 2003), but we do not know if they generalize to other host species and populations that are heavily parasitized by cowbirds. Where cowbirds have been culled in an attempt to protect endangered songbirds, the culls have only sometimes benefited the host population (Rothstein and Cook 2000). Since the prevalence and impact of cowbirds varies among hosts (e.g., Trine et al. 1998) a variety of demographic impacts are likely.

Immigration and Population Stability

Since relatively few song sparrows immigrated to Mandarte, the direct effect of immigration on the number of sparrows on the island was only small (chapter 4). However, immigration increased population size after episodes of catastrophic mortality and greatly increased the persistence of the population in viability models (chapter 10). A greater influence of immigration may explain the more stable dynamics observed in mainland song sparrow populations (Chase et al. 2005). However, strong philopatry is a widespread characteristic of song sparrows (Arcese 1989a; Arcese et al. 2002) and is not simply an atypical characteristic of birds inhabiting Mandarte. Some other songbirds are also strongly philopatric, and an unwillingness to disperse between neighboring islands contributes to the endangerment of the Seychelles warbler (Komdeur et al. 2004).

Relative Risks of Extirpation

We used population viability modeling to assess how catastrophes, immigration, and the effects of cowbirds affected the risk of extirpation for Mandarte's song sparrows (chapter 10). Catastrophic mortality posed the greatest risk. Removing the possibility of severe population crashes reduced the risk of extirpation by at least 10-fold. Isolating the population from immigrants and allowing the presence of brown-headed cowbirds increased extirpation risk to a smaller degree. However, the presence of cowbirds was predicted to reduce mean population size more than either catastrophes or an absence of immigrants (figure 10.4).

Using population viability analysis to assess the relative consequences of a range of scenarios is beginning to supercede the estimation of single population-specific values for extirpation probability (Reed et al. 2002). Our models suggest that facilitating immigration, by means of reserve design or deliberate population augmentation, can be a useful tactic for managers seeking to avoid population extirpation (Arcese 2003).

Generalizing from Small Populations on Islands

How far can demographic and genetic patterns observed in small island populations be generalized to mainland populations? The answer may be not very far. First, island populations can show different demography and dynamics from each other. For example, finches in Galápagos (Grant et al. 2000), Capricorn silvereye (McCallum et al. 2000), Soay sheep and red deer (Clutton-Brock and Coulson 2002), and song sparrow populations all show different trajectories, despite some similarities in the underlying processes.

Second, islands often differ from continental areas in environmental conditions. On small islands, the surrounding water moderates daily and seasonal variation in weather. There are often few competitors and predators on islands, and abundant nutrients are sometimes supplied by seabird guano (e.g., Harding et al. 2004). As a result of relaxed natural selection on islands, some island species adopt unusual lifestyles. For example, lizards frequently pollinate island plants but rarely do so on continents (Olesen and Valido 2003). Isolation probably reduces exposure to epidemic diseases, although the absence of acquired immunity may lead to high mortality when diseases do arrive (Grenfell and Harwood 1997).

Because island environments are often favorable, populations on small islands may rebound more rapidly from demographic bottlenecks than isolated populations on mainland areas. Variances in population growth rates, however, are larger in island populations (chapter 4), perhaps because environmental catastrophes are more frequent on islands.

In line with this view, environmental variance in population growth rate on Mandarte (0.41) is by far the highest of the few estimates available (see table 1.2 in Lande et al. 2003).

Despite these difficulties, and the emerging pattern that environmental variability is the principal risk for the extirpation of island populations, there remains room for optimism. The models of Sæther et al. (2000a) and Tufto et al. (2000) do an impressive job of describing the dynamics of Mandarte's song sparrow population based on annual variation in annual reproductive success (ARS) per individual, and without a specific knowledge of limiting mechanisms. Future studies may yet reveal deterministic patterns that reduce the apparent influence of extrinsic random effects. A similar understanding has been achieved for other bird populations (Sæther et al. 2002).

11.2. Genetic Risks to Small Populations

There are four principal genetic concerns associated with populations becoming small. First, inbreeding among close relatives may reduce average fitness. Second, genetic drift causes loss or fixation of alleles, thereby reducing genetic variation and the ability of the population to adapt to environmental change. Third, mutational load will accumulate in small populations because the efficacy of selection is reduced (e.g., Frankham et al. 2002). Fourth, if individuals are translocated to counteract genetic drift, populations may suffer outbreeding depression. Our study clearly illustrates the power of immigration to establish and maintain high levels of genetic variation (chapters 7, 8) in small populations. For example, as imagined in the extinction vortex, allelic diversity fell after the severe population crash of 1989, but it did not remain low. The trickle of immigrants to the Mandarte population restored lost allelic diversity within 3 years (chapter 7) and greatly altered the population's gene pool over 15 years (chapter 8).

Small populations of common species in mainland situations will often experience high levels of immigration and thus can remain genetically similar despite population bottlenecks. Endangered populations, however, will usually behave like distant island populations. If they are cut off from immigration, they can only regenerate lost allelic diversity by mutation (Groombridge et al. 2000).

Immigrants and their contributions to gene flow ameliorated one of the main genetic threats, loss of genetic variation, on Mandarte (chapters 7, 8). Inbreeding, on the other hand, was frequent on Mandarte, rising sharply after the most severe population bottleneck in 1989 despite immigration (chapter 7). Negative effects of close inbreeding are abundantly clear in the agricultural literature, and many similar results have now been found in natural populations (Keller and Waller 2002), including Mandarte Island. Considerable reductions in fitness were associated with inbreeding in song sparrows on Mandarte (chapter 7).

However, despite the occurrence of inbreeding depression, average fitness of song sparrows on Mandarte remained good (chapters 3, 4) at least partly because immigrants provided a constant influx of genetic variation (chapter 8; Keller et al. 2001). As a result, the high levels of inbreeding maintained through the 1990s were associated with decreased immigration compared to the previous decade (chapter 4), and the population failed to rebound after a further decline in numbers from 1997 through 1999 (chapters 4, 7). It remains uncertain whether inbreeding depression in juvenile survival of females or an increase in emigration amid declining habitat quality was responsible.

Inbreeding Depression in the Wild—A Red Herring?

What do our results say about the genetic risks of inbreeding in small populations? We showed that a bottleneck in population size increased levels of inbreeding on Mandarte (chapter 7). Results from other wild populations have confirmed that inbreeding depression occurs commonly in wild populations (Keller and Waller 2002). However, the debate over the occurrence of inbreeding depression in small wild populations (e.g., Caro and Laurenson 1994) was never particularly constructive. By focusing on whether or not inbreeding depression occurs in wild populations, conservation biologists were diverted from a more important issue. This issue is that the absence of inbreeding depression within a single population can indicate either a negligible genetic load or a genetic load that has become fixed (Keller and Waller 2002). The latter possibility, which is built into the extinction vortex, is the result that should worry conservation biologists. Indeed, the existence of inbreeding depression in a population can even be a good sign. If inbreeding depression is evident in a population, sufficient genetic variation is segregating that evolution remains possible and minimum fitness has not been fixed.

Thus, we need to know whether rates of inbreeding (i.e., the increase in average inbreeding per generation) typical for small populations of conservation concern are low enough for selection to prevent the fixation of deleterious genes (Frankel and Soulé 1981). It has long been known from the agricultural literature that selection for fertility and survival can only succeed if the increase in average inbreeding per generation is below 2–3%. If the increase in inbreeding exceeds 2–3%, selection cannot prevent the fixation of deleterious alleles and the average fitness of lines declines rapidly. The extinction vortex is the likely outcome of this process.

It will be interesting for future historians of science to investigate why late-twentieth-century conservation biologists focused so much on the magnitude of inbreeding depression rather than on the balance between inbreeding and selection. Two early books on conservation biology clearly identified the latter question as the crucial one (Soulé and Wilcox 1980; Frankel and Soulé 1981). Yet, we still know little about this balance in natural populations. One reason for this gap is that the populations where inbreeding depression has been measured in wild animals are not suited to studying the balance between selection and inbreeding: They are either too big or not isolated enough for allelic fixation to be likely (see, e.g., table 1 in Keller and Waller 2002). Our study on Mandarte illustrates the second point. The ecologically rare but genetically substantial rate of immigration (chapters 7, 8) keeps the rate of inbreeding per generation low.

To address the genetic threats in the extinction vortex, we need to focus on populations that behave like newly endangered populations, that is, show a substantial increase in average inbreeding each generation. All populations that receive immigrants (even if they arrive so infrequently that they are likely to go undetected) are poor model systems here. We suggest that replicated sets of small populations in a laboratory or field setting (e.g., Belovsky et al. 2002; England et al. 2003) will be fruitful for such studies. Even though statistical power to detect inbreeding effects may be low within each population, replication should yield reliable results.

We detected outbreeding depression in crosses between "native" Mandarte birds and immigrants (chapter 8). This result surprised us because we expected the gene flow generated by immigrants to restrict local adaptation. Outbreeding depression is expected only when distantly related populations are crossed or when there is considerable local geographic adaptation (Templeton 1986; Hughes et al. 2003). We did not expect either situation on Mandarte because surrounding islands provided the only immigrant of known origin, and provide similar ecological conditions. Perhaps other immigrants did come from distant populations. The outbreeding effect that we demonstrated on Mandarte, however, was subtle compared to other recent demonstrations (Ebert et al. 2002; Hughes et al. 2003) and may not have important practical consequences.

A final genetic concern is the erosion of genetic variation due to unequal reproductive success (chapter 9). The song sparrows of Mandarte Island, like many other populations that have been studied closely (e.g., case histories in Clutton-Brock 1988; Newton 1989a), exhibit marked skew in lifetime reproductive success. Most adult individuals rear few or no descendents, while some others are very successful. In any population, high levels of skew may decrease effective population size because alleles carried by unsuccessful individuals are lost. However, it is encouraging that reproductive skew was less pronounced on Mandarte when population size was small (chapter 9). In chapter 6, we showed that reproductive skew arising from extrapair matings did not reduce effective population size.

11.3. Integrating Demographic and Genetic Risks

In the opening quote in this chapter, Lande (1988a) emphasized the allure and difficulty of achieving an integrated understanding of genetic and demographic risks in nature. While we are still far from this goal, we have made good progress over the past 16 years.

First, the extinction vortex (chapter 1) has proven to be a useful conceptual model (e.g., p. 36 in Lande et al. 2003). We do, however, know that populations do not inevitably enter extinction vortices when they become small. Small populations often rebound after catastrophic losses (Newton 1998; see our chapter 4). Despite this knowledge, it remains prudent to assume that extinction vortices occur commonly and to heed the main message of this metaphor: wild populations should be kept large enough to resist the environmental, demographic, and genetic threats of living in small groups.

However, we have also learned that huge reductions in population size are to be expected from overharvesting (Fa et al. 2004), biological invasions (e.g., Sanders et al. 2003), epidemic diseases (e.g., Hochachka and Dhondt 2000), and extreme weather events (McCallum et al. 2000). As a result, the continued creation of new small populations is inevitable, and many of these will undoubtedly go extinct.

Second, when a species is reduced to a single small population, managers should first aim to increase its size by manipulating key demographic mechanisms identified by scientific study (Caughley 1994). Individuals may be moved to offshore islands, where the factors that caused the initial declines are absent or weak (Bell and Merton 2002). If irreversible damage to breeding habitat is the main threat, captive breeding may be needed to keep a species at least temporarily extant. Such "strong intervention" approaches will sometimes fail, but they have worked well with three endangered island birds, the Chatham Island robin (Butler and Merton 1992), the Seychelles warbler (Komdeur 1997), and the Mauritius kestrel (Jones et al. 1995). When managers have identified several such factors influencing populations, variance components for each might be analyzed and decomposed. Managers could then focus their attention on the factor that, if it were manipulated, would do most to ameliorate population declines (Arcese 2003).

A further area where recent progress has been made concerns the interaction between environment and inbreeding depression (Hanski and Gaggiotti 2004). In a serendipitous early example, Sewall Wright (1922) found that food quality greatly affected the magnitude of inbreeding depression observed in guinea pigs. Food quality featured in his experiments only because food scarcity in World War I forced him to reduce food quality in some inbred lines. In cactus finches in Galápagos, inbreeding depression is more severe when population density is high and food availability low (Keller et al. 2002). Effects of inbreeding in the endangered takahe (Jamieson et al. 2003) and in the song sparrow (chapter 7) vary with gender and age class, perhaps because of interactions between genes and the environment. In effect, when we manage both dispersal and population size, we may be minimizing both demographic and genetic risks.

In summary, while we have a long way to go to understand how far genetic and demographic mechanisms interact to affect population endangerment and recovery, the two factors may interact to depress population growth. While Lande was correct to emphasize demographic mechanisms in population recovery in 1988, recent work (e.g., Spielman et al. 2004) has confirmed that the early focus on genetic risks (Soulé and Wilcox 1980; Frankel and Soulé 1981) was also well founded. Both mechanisms can be strong, and interactions between them cannot be discounted.

11.4. Advice for Managers of Threatened and Endangered Species

The last 30 years have confirmed many of the concerns captured by the extinction vortex metaphor. The three principal concerns for managers thus remain preventing declines in existing populations, increasing the size of currently small populations, and establishing new populations where only one or two remnants exist. Tried and tested actions include supplementing food, improving habitat to increase fecundity, and controlling predators to reduce mortality rates (e.g., Komdeur 1996). Population viability models can rank key threats to well-described populations (e.g., chapter 10), but the necessary data will often be lacking. Augmenting the gene pool can reverse a negative population growth rate (e.g., Vilà et al. 2003), but translocating individuals that are too genetically distinct can do serious harm (e.g., Hughes et al. 2003). Managers must therefore choose carefully when assessing risks and benefits to a target population.

Cross-cultural differences in the willingness to use these methods exist. For example, managers of endangered species in New Zealand are quick to attack and kill presumed predators and competitors and to translocate populations. Europeans and North Americans are more reluctant to take radical action. It is relevant here that many predators in New Zealand and Australia are themselves introduced (Sinclair et al. 1998) and that habitats there are highly modified.

Optimism exists that landscape management and habitat restoration can help to set the table for threatened species to recover (e.g., Rosenzweig 2003). While landscape management may indeed be valuable, it will be costly and/or slow to provide results. As societies face increasing demands for conservation action, cost considerations will likely play an increasing role in all but the richest human societies.

Beyond these concerns are the likely effects of global climate change on populations (e.g., Sillett et al. 2000). If the more extreme predictions of changing climates are borne out (Houghton 2001; Green et al. 2001), the habitats of 2050 may not resemble today's habitats. It will then be wise to maintain viable source populations that are connected by dispersal. This may facilitate the action of selection on genetic variance in physiological traits related to climate tolerance. In specialized organisms, populations and species may have to be moved, but to where? Climate-based models and ambitious "seeding" experiments may help to predict where these places are before they are urgently needed.

11.5. Questions for Discussion

1. Should the view that most vertebrate populations are regulated be discarded in light of the recognition that environmental variability contributes substantially to varying population size?

2. Have conservation biologists been misled into focusing on the detection of inbreeding depression in natural populations?
3. Aggressive management of predators and competitors of threatened species is commonly practiced in New Zealand and Australia. Also, threatened populations are translocated frequently to predator-free sites. Should this "down under approach" be adopted more widely as the extinction crisis deepens?
4. Should hybridization be considered to rescue populations of a sexually reproducing species where individuals of only one sex remain?
5. How big a threat is climate change compared to the other factors threatening small populations, and how does the time scale considered affect your answer?
6. What new results of conservation interest could de derived from studying multiple populations of the same wild species in contrasting environments (e.g., Mougeot et al. 2003)?
7. Laboratory studies have been used to study the performance of small populations (e.g., Belovsky et al. 2002; England et al. 2003). Given that it is difficult to extrapolate across wild populations, how far can laboratory studies inform us of what will happen in small wild populations?

Appendix
Latin Names of Species of Plants and Animals Mentioned in Text

Common name	Latin name
Acorn woodpecker	*Melanerpes formicivorus*
Adder	*Vipera berus*
African hunting dog	*Lycaon pictus*
Arbutus	*Arbutus menziesii*
Australian freshwater shrimp	*Paratya australiensis*
Bald eagle	*Haliaeetus leucocephalus*
Barn owl	*Tyto alba*
Barn swallow	*Hirundo rustica*
Bay checkerspot butterfly	*Euphydryas editha*
Black-throated blue warbler	*Dendroica caerulescens*
Bighorn sheep	*Ovis canadensis*
Bison	*Bison bison*
Bitter cherry	*Prunus emarginata*
Black-capped chickadee	*Poecile atricapillus*
Blue grouse	*Dendragapus obscurus*
Blue tit	*Parus caeruleus*
Brown-headed cowbird	*Molothrus ater*
Cactus finch	*Geospiza scandens*
Camas	*Camassia quamash*
Canada goose	*Branta canadensis*
Cape Sable seaside sparrow	*Ammodramus maritimus mirabilis*
Capricorn silvereye	*Zosterops lateralis chlorocephala*
Caribou	*Rangifer tarandus*
Chatham Island black robin	*Petroica traversi*
Chokecherry	*Prunus virginiana*

Common name	Latin name
Coal tit	*Parus ater*
Collared flycatcher	*Ficedula albicollis*
Common cuckoo	*Cuculus canorus*
Common raven	*Corvus corax*
Cooper's hawk	*Accipiter cooperii*
Cougar	*Puma concolor*
Cutleaf blackberry	*Rubus laciniatus*
Deer mouse	*Peromyscus maniculatus*
Double-crested cormorant	*Phalacrocorax auritus*
Douglas fir	*Pseudotsuga menziesii*
Dunnock	*Prunella modularis*
Dusky seaside sparrow	*Ammodramous maritimus nigrescens*
European dipper	*Cinclus cinclus*
European robin	*Erithracus rubecula*
European sparrowhawk	*Accipiter nisus*
European starling	*Sturnus vulgaris*
Florida panther	*Puma concolor coryi*
Florida scrub-jay	*Aphelocoma coerulescens*
Fireweed	*Epilobium angustifolium*
Fox sparrow	*Passerella illiaca*
Fringecups	*Tellima grandiflora*
Fruit flies	*Drospohila melanogaster,*
	D. mercatorum
Galápagos mockingbird	*Nesomimus parvulus*
Garry oak	*Quercus garryana*
Glanville fritillary butterfly	*Melitaea cinxia*
Glaucous-winged gull	*Larus glaucescens*
Golden-crowned sparrow	*Zonotrichia atricapilla*
Grand fir	*Abies grandis*
Gray wolf	*Canis lupus*
Great horned owl	*Bubo virginianus*
Great reed warbler	*Acrocephalus arundinaceus*
Great tit	*Parus major*
Heath hen	*Tympanuchus cupido cupido*
Himalayan blackberry	*Rubus discolor*
House sparrow	*Passer domesticus*
Indian mynah	*Acridotheres tristis*
Ivy	*Hedera helix*
Kirtland's warbler	*Dendroica kirtlandii*
Land snail from Raitea (Society Islands)	*Partula turgida*
Large cactus finch	*Geospiza conirostris*
Mauritius kestrel	*Falco punctatus*
Meadow pipit	*Anthus pratensis*

Common name	Latin name
Medium ground finch	*Geospiza fortis*
Migratory locust	*Locusta migratoria*
Nettle	*Urtica dioica*
Nootka rose	*Rosa nutkana*
Northern elephant seal	*Mirounga angustirostris*
Northern gannet	*Morus bassanus*
Northern goshawk	*Accipiter gentilis*
Northern shrike	*Lanius excubitor*
Northwestern crow	*Corvus caurinus*
Ocean spray	*Holodiscus discolor*
Pelagic cormorant	*Phalacrocorax pelagicus*
Pied flycatcher	*Ficedula hypoleuca*
Pigeon guillemot	*Cepphus columba*
Platyfish	*Xiphophorus maculatus*
Purple martin	*Progne subis*
Red deer	*Cervus elaphus*
Red elderberry	*Sambucus racemosa*
Red-cockaded woodpecker	*Picoides borealis*
Red grouse	*Lagopus lagopus scoticus*
Red-winged blackbird	*Agelaius phoeniceus*
River otter	*Lutra canadensis*
Rock pipit	*Anthus petrosus*
Rufous-collared sparrow	*Zonotrichia capensis*
Saskatoon	*Amelanchier alnifolia*
Savannah sparrow	*Passerculus sandwichensis*
Seychelles magpie robin	*Copsychus sechellarum*
Seychelles warbler	*Acrocephalus sechellensis*
Shiny cowbird	*Molothrus bonariensis*
Snowberry	*Symphoricarpos albus*
Soay sheep	*Ovis aries*
Song sparrow	*Melospiza melodia*
Spotted towhee	*Pipilo maculatus*
Spruce grouse	*Falcipennis canadensis*
Squinting bush brown butterfly	*Bicyclus anynana*
Takahe	*Porphyrio hochstetteri*
Tree swallow	*Tachycineta bicolor*
White-crowned sparrow	*Zonotrichia leucophrys*
Water flea	*Daphnia magna*
Western red cedar	*Thuja plicata*
Willow	*Salix scouleriana*
Willow flycatcher	*Empidonax traillii*
Willow ptarmigan (red grouse)	*Lagopus lagopus*
Winter wren	*Troglodytes troglodytes*

REFERENCES

Adam, D., N. Dimitrijevic, and M. Schartl. 1993. Tumor suppression in *Xiphophorus* by an accidentally acquired promoter. Science 259:816–819.

Alberts, S. C., H. E. Watts, and J. Altmann. 2003. Queuing and queue-jumping: long-term patterns of reproductive skew in male savannah baboons, *Papio cynocephalus*. Animal Behaviour 65:821–840.

Albon, S. D., T. N. Coulson, D. Brown, F. E. Guinness, J. M. Pemberton, and T. H. Clutton-Brock. 2000. Temporal changes in key factors and key age groups influencing the population dynamics of female red deer. Journal of Animal Ecology 69:1099–1110.

Aldrich, J. W. 1984. Ecogeographical variation in size and proportions of the song sparrow (*Melospiza melodia*). Ornithological Monographs 35:1–134.

Allee, W. C. 1931. Animal aggregations: a study in general sociology. University of Chicago Press, Chicago, Illinois.

Allendorf, F. W., and N. Ryman. 2002. The role of genetics in population viability analysis. Pages 50–85 *in* S. R. Beissinger and D. R. McCullough, editors. Population viability analysis. University of Chicago Press, Chicago, Illinois.

Altum, B. 1868. Der Vogel und sein Leben. Münster.

Andersson, M. 1994. Sexual selection. Princeton University Press, Princeton, New Jersey.

Andrewartha, H. G., and L. C. Birch. 1954. The distribution and abundance of animals. University of Chicago Press, Chicago, Illinois.

Arcese, P. 1987. Age, intrusion pressure and defense against floaters by territorial male song sparrows. Animal Behaviour 35:773–784.

Arcese, P. 1989a. Intrasexual competition, mating system and natal dispersal in song sparrows. Animal Behaviour 37:958–979.

Arcese, P. 1989b. Territory acquisition and loss in male song sparrows. Animal Behaviour 37:45–55.

Arcese, P. 1989c. Intrasexual competition and the mating system in primarily monogamous birds: the case of the song sparrow. Animal Behaviour 37:96–111.

Arcese, P. 2003. Individual quality, environment, and conservation. Pages 271–297 in M. Festa-Bianchet and M. Apollonio, editors. Animal behavior and wildlife conservation. Island Press, Washinton, D.C.

Arcese, P., and J. N. M. Smith. 1985. Phenotypic correlates and ecological consequences of dominance in song sparrows. Journal of Animal Ecology 54:817–830.

Arcese, P., and J. N. M. Smith. 1988. Effects of population density and supplemental food on reproduction in song sparrows. Journal of Animal Ecology 57:119–136.

Arcese, P., and J. N. M. Smith. 1999. Impacts of nest depredation and brood parasitism on the productivity of North American passerines. Pages 2953–2966 (compact disc) in N. J. Adams and R. H. Slotow, editors. Proceedings of the 22nd International Ornithological Congress, Durban, South Africa.

Arcese, P., J. N. M. Smith, and M. I. Hatch. 1996. Nest predation by cowbirds and its consequences for passerine demography. Proceedings of the National Academy of Sciences of the United States of America 93:4608–4611.

Arcese, P., J. N. M. Smith, W. M. Hochachka, C. M. Rogers, and D. Ludwig. 1992. Stability, regulation, and the determination of abundance in an insular song sparrow population. Ecology 73:805–822.

Arcese, P., M. K. Sogge, A. B. Marr, and M. A. Patten. 2002. Song sparrow (Melospiza melodia). Pages 1–40 in A. F. Poole and F. B. Gill, editors. The birds of North America: life histories for the 21st century. No. 704. American Ornithologist's Union, Philadelphia, Pennsylvania.

Arcese, P., P. K. Stoddard, and S. M. Hiebert. 1988. The form and function of song in female song sparrows. Condor 90:44–50.

Armstrong, D. P., and J. G. Ewen. 2002. Dynamics and viability of a New Zealand robin population reintroduced to regenerating fragmented habitat. Conservation Biology 16:1074–1085.

Askenmo, C., and R. Neergaard. 1990. Polygyny and nest predation in the rock pipit. Do females trade male assistance against safety? Pages 331–343 in J. Blondel, A. Gosler, J. D. Lebreton, and R. McCleery, editors. Population biology of passerine birds. NATO Advanced Science Institutes Series No. 24. Springer-Verlag, Berlin.

Balshine, S., B. Leach, F. Neat, H. Reid, M. Taborsky, and N. Werner. 2001. Correlates of group size in a cooperatively breeding cichlid fish (Neolamprologus pulcher). Behavioral Ecology and Sociobiology 50:134–140.

Bard, S. C., M. Hau, M. Wikelski, and J. C. Wingfield. 2002. Vocal distinctiveness and response to conspecific playback in the spotted antbird, a Neotropical suboscine. Condor 104:387–394.

Beecher, M. D., S. E. Campbell, and J. C. Nordby. 2000. Territory tenure in song sparrows is related to song sharing with neighbours, but not to repertoire size. Animal Behaviour 59:29–37.

Beissinger, S. R. 2002. Population viability analysis: past, present and future. Pages 5–17 in S. R. Beissinger and D. R. McCullough, editors. Population viability analysis. Chicago University Press, Chicago, Illinois.

Beissinger, S. R., and D. R. McCullough, editors. 2002. Population viability analysis. University of Chicago Press, Chicago, Illinois.

Beissinger, S. R., and M. I. Westphal. 1998. On the use of demographic models of population viability in endangered species management. Journal of Wildlife Management 62:821–841.

Beletsky, L. D., and G. H. Orians. 1996. Red-winged blackbirds. University of Chicago Press, Chicago, Illinois.

Bell, B. D., and D. V. Merton. 2002. Critically endangered bird populations and their management. Pages 105–138 in K. Norris and D. J. Pain, editors. Conserving bird biodiversity: General principles and their application. Cambridge University Press, Cambridge.

Belovsky, G. E., C. Mellison, C. Larson, and P. A. van Zandt. 2002. How good are PVA models? Testing their predictions with experimental data on brine shrimp. Pages 257–283 in S. R. Beissinger and D. R. McCullough, editors. Population viability analysis. Chicago University Press, Chicago, Illinois.

Bent, A. C. 1932. Life histories of North American gallinaceous birds. Smithsonian Institution and United States National Museum Bulletin No. 162, Washington, D.C.

Benton, T. G., and A. Grant. 2000. Evolutionary fitness in ecology: comparing measures of fitness in stochastic, density-dependent environments. Evolutionary Ecology Research 2:769–789.

Benton, T. G., E. Ranta, V. Kaitala, and A. P. Beckerman. 2001. Maternal effects and the stability of population dynamics in noisy environments. Journal of Animal Ecology 70:590–599.

Berger, J. 1990. Persistence of different-sized populations: an empirical-assessment of rapid extinctions in bighorn sheep. Conservation Biology 4:91–98.

Berger, L., R. Speare, P. Daszak, D. E. Green, A. A. Cunningham, C. L. Goggin, et al. 1998. Chytridiomycosis causes amphibian mortality associated with population declines in the rain forests of Australia and Central America. Proceedings of the National Academy of Sciences of the United States of America 95:9031–9036.

BirdLife International. 2000. Threatened birds of the world. Lynx Edicions, Barcelona, Spain; BirdLife International, Cambridge.

Birkhead, T. R., and A. P. Møller. 1992. Sperm competition in birds: evolutionary causes and consequences. Academic Press, London.

Black, J. M., editor. 1996. Partnerships in birds: the study of monogamy. Oxford University Press, Oxford.

Blackburn, T. M., P. Cassey, R. P. Duncan, K. L. Evans, and K. J. Gaston. 2004. Avian extinction and mammalian introductions on oceanic islands. Science 305:1955–1958.

Boag, D. A., and M. A. Schroeder. 1987. Population fluctuations in spruce grouse. What determines their numbers in spring? Canadian Journal of Zoology 65:2430–2435.

Both, C. 2000. Experimental evidence for density dependence in great tits. Journal of Animal Ecology 67:667–674.

Bourke, A. F. G., and J. Heinze. 1994. The ecology of communal breeding—the case of multiple-queen leptothoracine ants. Philosophical Transactions of the Royal Society of London B 345:359–372.

Bower, J. L. 2000. Acoustic interactions during naturally occurring territorial conflict in a song sparrow neighborhood. Ph.D. Thesis. Cornell University, Ithaca, New York.

Boyce, M. S. 1992. Population viability analysis. Annual Review of Ecology and Systematics 23:481–506.

Boyce, M. S. 2002. Reconciling the small-population and declining-population paradigms. Pages 41–49 in S. R. Beissinger and D. R. McCullough, editors. Population viability analysis. University of Chicago Press, Chicago, Illinois.

Breininger, D. R., V. L. Larson, D. M. Oddy, R. B. Smith, and M. J. Barkaszi. 1996. Florida scrub jay demography in different landscapes. Auk 113:617–625.

Briskie, J. V., and M. Mackintosh. 2004. Hatching failure increases with severity of population bottlenecks in birds. Proceedings of the National Academy of Sciences of the United States of America 101:558–561.

Briskie, J. V., and S. G. Sealy. 1990. Evolution of short incubation periods in the parasitic cowbirds Molothrus spp. Auk 107:789–794.

Brittingham, M. C., and S. A. Temple. 1983. Have cowbirds caused forest songbirds to decline? BioScience 33:31–35.

Brommer, J. E., J. Merilä, and H. Kokko. 2002. Reproductive timing and individual fitness. Ecology Letters 5:802–810.

Brook, B. W., and J. Kikkawa. 1998. Examining threats faced by island birds: a population viability analysis on the Capricorn silvereye using longterm data. Journal of Applied Ecology 35:491–503.

Brooker, L., and M. Brooker. 2002. Dispersal and population dynamics of the blue-breasted fairy wren, Malurus pulcherrimus, in fragmented habitat in the Western Australian wheatbelt. Wildlife Research 29:225–233.

Brown, J. H., and A. Kodric-Brown. 1977. Turnover rates in insular biogeography: effect of immigration on extinction. Ecology 58:445–449.

Brown, J. L. 1964. The integration of agonistic behavior in the Steller's jay *Cyanocitta stelleri*. University of California Publications in Zoology 60:223–328.

Brown, J. L. 1969. Territorial behavior and population regulation in birds: a review and re-evaluation. Wilson Bulletin 81:293–329.

Burke, D. M., and E. Nol. 1998. Influence of food abundance, nest-site habitat, and forest fragmentation on breeding ovenbirds. Auk 115:96–104.

Burnham, K. P., and D. R. Anderson. 2002. Model selection and multi-model inference, 2nd ed. Springer, New York.

Burton, R. S., P. D. Rawson, and S. Edmands. 1999. Genetic architecture of physiological phenotypes: empirical evidence for coadapted gene complexes. American Zoologist 39:451–462.

Butler, D., and D. Merton. 1992. The Black robin: saving the world's most endangered bird. Oxford University Press, Auckland, New Zealand.

Butler, R. W., N. A. M. Verbeek, and H. Richardson. 1984. The breeding biology of the northwestern crow. Wilson Bulletin 96:408–418.

Byers, C., J. Curson, and U. Olsson. 1995. Sparrows and buntings. A guide to the sparrows and buntings of North America and the world. Houghton Mifflin Company, Boston, Massachusetts.

Cam, E., J. Y. Monnat, and J. E. Hines. 2003. Long-term fitness consequences of early conditions in the kittiwake. Journal of Animal Ecology 72:411–424.

Cann, R. L., and L. J. Douglas. 1999. Parasites and conservation of Hawaiian birds. Pages 121–136 *in* L. F. Landweber and A. P. Dobson, editors. Genetics and the extinction of species. Princeton University Press, Princeton, New Jersey.

Caro, T. M., and M. K. Laurenson. 1994. Ecological and genetic factors in conservation: a cautionary tale. Science 263:485–486.

Cassidy, A. L. E. V. 1993. Song variation and learning in island populations of song sparrows. Ph.D. Thesis. University of British Columbia, Vancouver.

Cassirer, E. F. 2005. Ecology of disease in bighorn sheep in Hell's Canyon USA. Ph.D. thesis. University of British Columbia, Vancouver.

Catchpole, C. K., and P. J. B. Slater. 1995. Bird song: biological themes and variations. Cambridge University Press, Cambridge.

Catterall, C. P., W. S. Wyatt, and L. J. Henderson. 1982. Food resources, territory density and reproductive success of an island silvereye population *Zosterops lateralis*. Ibis 124:405–421.

Caughley, G. 1994. Directions in conservation biology. Journal of Animal Ecology 63:215–244.

Caughley, G., and A. Gunn. 1993. Dynamics of large herbivores in deserts: kangaroos and caribou. Oikos 67:47–55.

Caughley, G., and A. Gunn. 1996. Conservation biology in theory and practice. Blackwell Science, Cambridge, Massachusetts.

Cavalli-Sforza, L. L., and W. F. Bodmer. 1971. The genetics of human populations. Freeman, San Francisco, California.

Chan, Y., and P. Arcese. 2002. Subspecific differentiation and conservation of song sparrows (*Melospiza melodia*) in the San Francisco Bay region inferred by microsatellite loci analysis. Auk 119:641–657.

Chan, Y., and P. Arcese. 2003. Morphological and microsatellite differentiation in *Melospiza melodia* (Aves) at a microgeographic scale. Journal of Evolutionary Biology 16:939–947.

Chance, E. P. 1940. The truth about the cuckoo. Country Life, London.

Charlesworth, B., and D. Charlesworth. 1999. The genetic basis of inbreeding depression. Genetical Research 74:329–340.

Chase, M. K., N. Nur, and G. R. Geupel. 2005. Effects of weather and population density on reproductive success and population dynamics in a song sparrow population: a long-term study. Auk 122: 571–592.

Cheptou, P. O., A. Berger, A. Blanchard, C. Collin, and J. Escarre. 2000. The effect of drought stress on inbreeding depression in four populations of the Mediterranean outcrossing plant *Crepis sancta* (Asteraceae). Heredity 85:294–302.

Chitty, D. 1967. The natural selection of self-regulatory behaviour in animal populations. Proceedings of the Ecological Society of Australia 2:51–78.

Clobert, J., C. M. Perrins, R. H. McCleery, and A. G. Gosler. 1988. Survival rate in the great tit *Parus major* in relation to sex, age, and immigration status. Journal of Animal Ecology 57:287–306.

Clutton-Brock, T. H., editor. 1988. Reproductive success: studies of individual variation in contrasting breeding systems. University of Chicago Press, Chicago, Illinois.

Clutton-Brock, T. H., and T. Coulson. 2002. Comparative ungulate dynamics: the devil is in the detail. Philosophical Transactions of the Royal Society of London B 357:1285–1298.

Clutton-Brock, T. H., and J. M. Pemberton, editors. 2004. Soay sheep. Cambridge University Press, Cambridge.

Cohen, J. E. 2003. Human population: the next half-century. Science 302: 1172–1175.

Cole, L. C. 1954. The population consequences of life history phenomena. Quarterly Review of Biology 29:103–137.

Coltman, D. W., J. A. Smith, D. R. Bancroft, J. G. Pilkington, A. D. C. MacColl, T. H. Clutton-Brock, and J. M. Pemberton. 1999. Density-dependent variation in lifetime breeding success and natural and sexual selection in Soay rams. American Naturalist 154:730–746.

Cooney, R., and N. C. Bennett. 2000. Inbreeding avoidance and reproductive skew in a cooperative mammal. Proceedings of the Royal Society of London B 267:801–806.

Coulson, T., E. A. Catchpole, S. D. Albon, B. J. T. Morgan, J. M. Pemberton, T. H. Clutton-Brock, et al. 2001. Age, sex, density, winter weather, and population crashes in Soay sheep. Science 292:1528–1531.

Courchamp, F., and D. B. Macdonald. 2001. Crucial importance of pack size in the African wild dog *Lycaon pictus*. Animal Conservation 4:169–174.

Cramp, S., editor. 1988. Handbook of the birds of Europe, the Middle East, and North Africa. The birds of the western palaearctic. Vol. 5: tyrant flycatchers to thrushes. Oxford University Press, Oxford.

Cramp, S., editor. 1992. Handbook of the birds of Europe, the Middle East, and North Africa. The birds of the western palaearctic. Vol. 6: warblers. Oxford University Press, Oxford.

Cramp, S., and C. M. Perrins, editors. 1993. Handbook of the birds of Europe, the Middle East, and North Africa. The birds of the western palaearctic. Vol. 7: flycatchers to shrikes. Oxford University Press, Oxford.

Cramp, S., and C. M. Perrins, editors. 1994a. Handbook of the birds of Europe, the Middle East, and North Africa. The birds of the western palaearctic. Vol. 8: crows to finches. Oxford University Press, Oxford.

Cramp, S., and C. M. Perrins, editors. 1994b. Handbook of the birds of Europe, the Middle East, and North Africa. The birds of the western palaearctic. Vol. 9: buntings and New World warblers. Oxford University Press, Oxford.

Creel, S., and N. M. Creel. 2002. The African wild dog: behavior, ecology, and conservation. Princeton University Press, Princeton, New Jersey.

Crnokrak, P., and S. C. H. Barrett. 2002. Perspective: purging the genetic load: a review of the experimental evidence. Evolution 56:2347–2358.

Crnokrak, P., and D. A. Roff. 1999. Inbreeding depression in the wild. Heredity 83:260–270.

Crow, J. F. 1948. Alternative hypotheses of hybrid vigor. Genetics 33:477–487.

Crow, J. F. 1952. Dominance and overdominance. Pages 282–297 *in* J. W. Gowen, editor. Heterosis: a record of researches directed toward explaining and utilizing the vigor of hybrids. Iowa State College Press, Ames, Iowa.

Crow, J. F., and M. Kimura. 1970. An introduction to population genetics theory. Alpha Editions, Minneapolis, Minnesota.

Cunningham, A. A. 1996. Disease risks of wildlife translocations. Conservation Biology 10:349–353.

Cunningham, A. A., and P. Daszak. 1998. Extinction of a species of land snail due to infection with a microsporidian parasite. Conservation Biology 12:1139–1141.

Curry, R. L., and P. R. Grant. 1989. Demography of the cooperatively breeding Galápagos mockingbird, *Nesomimus parvulus*, in a climatically variable environment. Journal of Animal Ecology 58:441–463.

Curson, D. R., C. B. Goguen, and N. E. Mathews. 2000. Long-distance commuting by brown-headed cowbirds in New Mexico. Auk 117:795–799.

Dahlgaard, J., and A. A. Hoffmann. 2000. Stress resistance and environmental dependency of inbreeding depression in *Drosophila melanogaster.* Conservation Biology 14:1187–1192.

Darwin, C. 1868. The variation of animals and plants under domestication. J. Murray and Co., London.

Darwin, C. 1876. The effects of cross and self fertilization in the vegetable kingdom. J. Murray and Co., London.

Davies, N. B. 1992. Dunnock behaviour and social evolution. Oxford University Press, Oxford.

Davies, N. B. 2000. Cuckoos, cowbirds and other cheats. T. and A. D. Poyser, Ltd., London.

Davies, N. B., and A. Lundberg. 1984. Food distribution and a variable mating system in the dunnock *Prunella modularis.* Journal of Animal Ecology 53:895–912.

DeCapita, M. E. 2000. Brown-headed cowbird control on Kirtland's warbler nesting areas in Michigan, 1972–1995. Pages 333–341 *in* J. N. M. Smith, T. L. Cook, S. I. Rothstein, S. K. Robinson, and S. G. Sealy, editors. Ecology and management of cowbirds and their hosts. University of Texas Press, Austin, Texas.

De Kogel, C. H. 1997. Long-term effects of brood size manipulation on morphological development and sex-specific mortality of offspring. Journal of Animal Ecology 66:167–178.

Denno, R. F., and M. A. Peterson. 1995. Density-dependent dispersal and its consequences for population dynamics. Pages 113–130 *in* N. Cappuccino and P. W. Price, editors. Population dynamics: new approaches and synthesis. Academic Press, San Diego, California.

Désrochers, A., and R. D. Magrath. 1993. Age-specific fecundity in European blackbirds (*Turdus merula*)—individual and population trends. Auk 110:255–263.

Désrochers, A., and R. D. Magrath. 1996. Divorce in the European blackbird: seeking greener pastures? Pages 177–191 *in* J. M. Black, editor. Partnerships in birds: the study of monogamy. Oxford University Press, Oxford.

Dhondt, A. A. 1979. Summer dispersal and survival of juvenile great tits in southern Sweden. Oecologia 42:139–157.

Dhondt, A. A., F. Adriaensen, and W. Plompen. 1996. Between and within-population variability in the great tit. Pages 235–248 *in* J. M. Black, editor. Partnerships in birds: the study of monogamy. Oxford University Press, Oxford.

Diamond, J. M. 1984. "Normal" extinctions of isolated populations. Pages 191–245 *in* M. H. Nitecki, editor. Extinctions. University of Chicago Press, Chicago, Illinois.

Drent, P. J. 1984. The functional ethology of territoriality in the great tit (*Parus major*). Ph.D. Thesis. Zoological Laboratory, University of Groningen, The Netherlands.

Drent, P. J., G. F. van Tets, F. S. Tompa, and K. Vermeer. 1964. The breeding birds of Mandarte Island, British Columbia. The Canadian Field-Naturalist 78:208–263.

Eberhardt, L. L. 1988. Using age structure data from changing populations. Journal of Applied Ecology 25:373–378.

Ebert, D., C. Haag, M. Kirkpatrick, M. Riek, J. W. Hottinger, and V. I. Pajunen. 2002. A selective advantage to immigrant genes in a *Daphnia* metapopulation. Science 295:485–488.

Edmands, S. 2002. Does parental divergence predict reproductive compatibility? Trends in Ecology and Evolution 17:520–527.

Ehrlich, P. R., D. S. Dobkin, and D. Wheye. 1988. The birder's handbook. A field guide to the natural history of North American birds. Simon and Schuster, New York.

Ehrlich, P. R., and I. Hanski, editors. 2004. On the wings of checkerspots—a model system for population biology. Oxford University Press, Oxford.

Ekman, J., S. Eggers, and M. Griesser. 2002. Fighting to stay: the role of sibling rivalry for delayed dispersal. Animal Behaviour 64:453–459.

Ellner, S. P., J. Fieberg, D. Ludwig, and C. Wilcox. 2002. Precision of population viability analysis. Conservation Biology 16:258–261.

Elphick, C., J. B. Dunning, Jr., and D. A. Sibley. 2001. The Sibley guide to bird life and behavior. Chanticleer Press, A. A. Knopf, Inc., New York.

Emlen, S. T., and L. W. Oring. 1977. Ecology, sexual selection, and evolution of mating systems. Science 197:215–223.

Engen, S., R. Lande, and B. E. Sæther. 2003. Demographic stochasticity and Allee effects in populations with two sexes. Ecology 84:2378–2386.

Engen, S., and B. E. Sæther. 2000. Predicting the time to quasi-extinction for populations far below their carrying capacity. Journal of Theoretical Biology 205:649–658.

England, P. R., G. H. R. Osler, L. M. Woodworth, M. E. Montgomery, D. A. Briscoe, and R. Frankham. 2003. Effects of intense versus diffuse population bottlenecks on microsatellite genetic diversity and evolutionary potential. Conservation Genetics 4:595–604.

Ens, B. J., S. Choudhury, and J. M. Black. 1996. Mate fidelity and divorce in monogamous birds. Pages 344–395 *in* J. M. Black, editor. Partnerships in birds: the study of monogamy. Oxford University Press, Oxford.

Fa, J. E., S. F. Ryan, and D. J. Bell. 2004. Hunting vulnerability, ecological characteristics and harvest rates of bushmeat species in Afrotropical forests. Biological Conservation 121:167–176.

Falconer, D. S., and T. F. C. Mackay. 1996. Introduction to quantitative genetics, 4th ed. Longman Group Ltd., Essex, England.

Fisher, R. A. 1930. The genetical theory of natural selection. Clarendon Press, Oxford.

Forstmeier, W., D. P. J. Kuijper, and B. Leisler. 2001. Polygyny in the dusky warbler, *Phylloscopus fuscatus*: the importance of female qualities. Animal Behaviour 62:1097–1108.

Forstmeier, W., and B. Leisler. 2004. Repertoire size, sexual selection, and offspring viability in the great reed warbler: changing patterns in space and time. Behavioral Ecology 15:555–563.

Frankel, O. H. 1974. Genetic conservation: Our evolutionary responsibility. Genetics 78:53–65.

Frankel, O. H., and M. E. Soulé. 1981. Conservation and evolution. Cambridge University Press, Cambridge.

Frankham, R. 1998. Inbreeding and extinction: a threshold effect. Conservation Biology 9:792–799.

Frankham, R., J. D. Ballou, and D. A. Briscoe. 2002. Introduction to conservation genetics. Cambridge University Press, Cambridge.

Frankham, R., K. Lees, M. E. Montgomery, P. R. England, E. Lowe, and D. A. Briscoe. 1999. Do population bottlenecks reduce evolutionary potential? Animal Conservation 2:255–260.

Franklin, A. B., D. R. Anderson, R. J. Gutierrez, and K. P. Burnham. 2000. Climate, habitat quality, and fitness in northern spotted owl populations in northwestern California. Ecological Monographs 70:539–590.

Franklin, I. R. 1980. Evolutionary change in small populations. Pages 135–150 *in* M. E. Soulé and B. A. Wilcox, editors. Conservation biology: an evolutionary-ecological perspective. Sinauer Associates, Sunderland, Massachusetts.

Franzreb, K. E., and K. V. Rosenburg. 1997. Are forest songbirds declining? Status assessment from the southern Appalachian and northeastern forests. Transactions of the North American Wildlife and Natural Resource Conference 62:264–279.

Freed, L. A. 1999. Extinction and endangerment of Hawaiian honeycreepers: a comparative approach. Pages 137–162 *in* L. F. Landweber and A. P. Dobson, editors. Genetics and the extinction of species. Princeton University Press, Princeton, New Jersey.

Friedmann, H. 1929. The cowbirds. A study in the biology of social parasitism. Charles C. Thomas, Springfield, Illinois.

Friedmann, H., and L. F. Kiff. 1985. The parasitic cowbirds and their hosts. Proceedings of the Western Foundation of Vertebrate Zoology 2:226–304.

Fry, A. J., and R. M. Zink. 1998. Geographic analysis of nucleotide diversity and song sparrow (Aves: Emberizidae) population history. Molecular Ecology 7:1303–1313.

Gaggiotti, O. E. 2003. Genetic threats to population persistence. Annales Zoologici Fennici 40:155–168.

Gaillard, J. M., M. Festa-Bianchet, N. G. Yoccoz, A. Loison, and C. Toïgo. 2000. Temporal variation in fitness components and population dynamics of large herbivores. Annual Review of Ecology and Systematics 31:367–393.

Gaines, M. S., and L. R. McClenaghan. 1980. Dispersal in small mammals. Annual Review of Ecology and Systematics 11:163–196.

Gaston, K. J., T. M. Blackburn, and K. K. Gooldwijk. 2003. Habitat conversion and global avian biodiversity loss. Proceedings of the Royal Society of London B 270:1293–1300.

Ghalambor, C. K., and T. E. Martin. 2001. Fecundity-survival trade-offs and parental risk-taking in birds. Science 292:494–497.

Gill, F. B. 1994. Ornithology, 2nd ed. Freeman and Company, New York.

Gilliard, E. T. 1969. Birds of paradise and bowerbirds. Natural History Press, Garden City, New Jersey.

Gilpin, M. E., and M. E. Soulé. 1986. Minimum viable populations: processes of extinction. Pages 19–34 *in* M. E. Soulé, editor. Conservation biology, the science of scarcity and diversity. Sinauer Assiocates, Sunderland, Massachusetts.

Glémin, S. 2003. How are deleterious mutations purged? Drift versus nonrandom mating. Evolution 57:2678–2687.

Goodman, D. 1987. The demography of chance extinction. Pages 11–34 *in* M. E. Soulé, editor. Viable populations for conservation. Cambridge University Press, Cambridge.

Granfors, D. A., P. J. Pietz, and L. A. Joyal. 2001. Frequency of egg and nestling destruction by female brown-headed cowbirds at grassland nests. Auk 118:765–769.

Grant, B. R., and P. R. Grant. 1989. Evolutionary dynamics of a natural population. The large cactus finch of the Galápagos. Princeton University Press, Princeton, New Jersey.

Grant, B. R., and P. R. Grant. 1996a. Cultural inheritance of song and its role in the evolution of Darwin's finches. Evolution 50:2471–2487.

Grant, P. R., and B. R. Grant. 1996b. Finch communities in a climatically fluctuating environment. Pages 343–390 *in* M. L. Cody and J. A. Smallwood, editors. Long-term studies of vertebrate communities. Academic Press, New York.

Grant, P. R., B. R. Grant, L. F. Keller, and K. Petren. 2000. Effects of El Niño events on Darwin's finch productivity. Ecology 81:2442–2457.

Grant, P. R., B. R. Grant, and K. Petren. 2001. A population founded by a single pair of individuals: establishment, expansion, and evolution. Genetica 112:359–382.

Green, R. E., M. Harley, M. Spalding, and C. Zockler, editors. 2001. Impacts of climate change on wildlife. Royal Society for the Preservation of Birds, Sandy, Bredfordshire, UK.

Greenwood, J. J. D., and S. R. Baillie. 1991. Effects of density-dependence and weather on population changes of English passerines using a non-experimental paradigm. Ibis 133:121–133.

Greenwood, P. J. 1980. Mating systems, philopatry and dispersal in birds and mammals. Animal Behaviour 28:1140–1162.

Grenfell, B., and J. Harwood. 1997. (Meta)population dynamics of infectious diseases. Trends in Ecology and Evolution 12:395–399.

Griffith, J. T., and J. C. Griffith. 2000. Cowbird control and the endangered least Bell's vireo: a management success story. Pages 342–356 *in* J. N. M. Smith, T. L. Cook, S. I. Rothstein, S. K. Robinson, and S. G. Sealy, editors. Ecology and management of cowbirds and their hosts. University of Texas Press, Austin, Texas.

Griffith, S. C. 2000. High fidelity on islands: a comparative study of extra-pair paternity in passerine birds. Behavioral Ecology 11:265–273.

Grønstøl, G. B., I. Byrkjedal, and Ø. Fiksen. 2003. Predicting polygynous settlement while incorporating varying female competitive strength. Behavioral Ecology 14:257–267.

Groom, M. J. 1998. Allee effects limit population viability of an annual plant. American Naturalist 151:487–496.

Groombridge, J. J., M. W. Bruford, C. G. Jones, and R. A. Nichols. 2001. Evaluating the severity of the population bottleneck in the Mauritius kestrel *Falco punctatus* from ringing records using MCMC estimation. Journal of Animal Ecology 70:401–409.

Groombridge, J. J., C. G. Jones, M. W. Bruford, and R. A. Nichols. 2000. Conservation biology—'ghost' alleles of the Mauritius kestrel. Nature 403:616.

Guerrant, E. O., Jr. 1992. Genetic and demographic considerations in the sampling of rare plants. Pages 321–344 *in* P. L. Fiedler and S. K. Jain, editors. Conservation biology: the theory and practice of nature conservation, preservation and management. Chapman and Hall, New York.

Gustafsson, L., and W. J. Sutherland. 1988. The costs of reproduction in the collared flycatcher *Ficedula albicollis*. Nature 335:813–815.

Hannon, S. J. 1983. Spacing and breeding density of willow ptarmigan in response to an experimental alteration of sex ratio. Journal of Animal Ecology 52:807–820.

Hanski, I. 1999. Metapopulation ecology. Oxford University Press, Oxford.

Hanski, I. 2002. Metapopulations of animals in high fragmented landscapes and population viability analysis. Pages 86–108 *in* S. R. Beissinger and D. R. McCullough, editors. Population viability analysis. Chicago University Press, Chicago, Illinois.

Hanski, I., C. J. Breuker, K. Schops, R. Setchfield, and M. Nieminen. 2002. Population history and life history influence the migration rate of female Glanville fritillary butterflies. Oikos 98:87–97.

Hanski, I., and O. E. Gaggiotti, editors. 2004. Ecology, genetics and evolution of metapopulations. Elsevier Academic Press, Amsterdam.

Harcourt, S. 1991. Endangered species. Nature 354:10.

Harding, J. S., D. J. Hawke, R. N. Holdaway, and M. J. Winterbourn. 2004. Incorporation of marine-derived nutrients from petrel breeding colonies into stream food webs. Freshwater Biology 49:576–586.

Hare, M. P., and G. F. Shields. 1992. Mitochondrial-DNA variation in the polytypic Alaskan song sparrow. Auk 109:126–132.

Harper, D. G. C. 1985. Brood division in robins. Animal Behaviour 33:466–480.

Harrison, S., and C. Ray. 2002. Plant population viability and metapopulation-level processes. Pages 109–122 *in* S. R. Beissinger and D. R. McCullough, editors. Population viability analysis. Chicago University Press, Chicago, Illinois.

Harvey, P. H., M. J. Stenning, and B. Campbell. 1988. Factors influencing reproductive success in the pied flycatcher. Pages 189–200 *in* T. H. Clutton-Brock, editor. Reproductive success. University of Chicago Press, Chicago, Illinois.

Hasselquist, D., S. Bensch, and T. von Schantz. 1996. Correlation between male song repertoire, extra-pair paternity and offspring survival in the great reed warbler. Nature 381:229–232.

Hasselquist, D., and P. W. Sherman. 2001. Social mating systems and extra-pair fertilizations in passerine birds. Behavioral Ecology 12:457–466.

Hauber, M. E. 2000. Nest predation and cowbird parasitism in song sparrows. Journal of Field Ornithology 71:389–398.

Hauber, M. E. 2002. Is reduced clutch size a cost of parental care in eastern phoebes (*Sayornis phoebe*)? Behavioral Ecology and Sociobiology 51:503–509.

Hauber, M. E., and S. A. Russo. 2000. Perch proximity correlates with higher rates of cowbird parasitism of ground nesting song sparrows. Wilson Bulletin 112:150–153.

Haydock, J., and W. D. Koenig. 2002. Reproductive skew in the polygynandrous acorn woodpecker. Proceedings of the National Academy of Sciences of the United States of America 99:7178–7183.

Hedrick, P. W. 1994. Purging inbreeding depression and the probability of extinction: full-sib mating. Heredity 73:363–372.

Hedrick, P. W., and S. T. Kalinowski. 2000. Inbreeding depression in conservation biology. Annual Review of Ecology and Systematics 31:139–162.

Hedrick, P. W., R. C. Lacy, F. W. Allendorf, and M. E. Soulé. 1996. Directions in conservation biology: comments on Caughley. Conservation Biology 10:1312–1320.

Heisey, D. M. 1992. Proportional hazards analysis of left-truncated survival data. SAS Institute Inc. Pages 1271–1276 *in* Proceedings of the 17th annual SAS user's group international conference. SAS Institute Inc., Cary, North Carolina.

Henry, P. Y., R. Pradel, and P. Jarne. 2003. Environment-dependent inbreeding depression in a hermaphroditic freshwater snail. Journal of Evolutionary Biology 16:1211–1222.

Hensley, M. M., and J. B. Cope. 1951. Further data on removal and repopulation of the breeding birds in a spruce-fir forest community. Auk 68:483–493.

Herkert, J. R., D. L. Reinking, D. A. Wiedenfeld, M. Winter, J. L. Zimmerman, W. E. Jensen, et al. 2003. Effects of prairie fragmentation on the nest success of breeding birds in the midcontinental United States. Conservation Biology 17:587–594.

Hiebert, S. M., P. K. Stoddard, and P. Arcese. 1989. Repertoire size, territory acquisition and reproductive success in the song sparrow. Animal Behaviour 37:266–273.

Hilborn, R., and M. Mangel. 1997. The ecological detective: confronting models with data. Princeton University Press, Princeton, New Jersey.

Hill, C. E., S. E. Campbell, J. C. Nordby, J. M. Burt, and M. D. Beecher. 1999. Song sharing in two populations of song sparrows (*Melospiza melodia*). Behavioral Ecology and Sociobiology 46:341–349.

Hobday, A. J., M. J. Tegner, and P. L. Haaker. 2001. Overexploitation of a broadcast spawning marine invertebrate: decline of the white abalone. Reviews in Fish Biology and Fisheries 10:493–514.

Hochachka, W. M. 1990. Seasonal decline in reproductive performance of song sparrows. Ecology 71:1279–1288.

Hochachka, W. M., and A. A. Dhondt. 2000. Density-dependent decline of host abundance resulting from a new infectious disease. Proceedings of the National Academy of Sciences of the United States of America 97:5303–5306.

Hochachka, W. M., and J. N. M. Smith. 1991. Determinants and consequences of nestling condition in song sparrows. Journal of Animal Ecology 60:995–1008.

Hochachka, W. M., J. N. M. Smith, and P. Arcese. 1989. Song sparrow. Pages 135–152 *in* I. Newton, editor. Lifetime reproduction in birds. Academic Press, London.

Hodges, K. E., C. J. Krebs, D. S. Hik, C. I. Stefan, E. A. Gillis, and C. E. Doyle. 2000. Snowshoe hare demography. Pages 141–178 *in* C. J. Krebs, S. Boutin, and R. Boonstra, editors. Ecosystem dynamics of the boreal forest. Oxford University Press, Oxford.

Hoelzel, A. R., B. J. Le Boeuf, J. Reiter, and C. Campagna. 1999. Alpha-male paternity in elephant seals. Behavioral Ecology and Sociobiology 46:298–306.

Hoenig, J. M., and D. M. Heisey. 2001. The abuse of power: the pervasive fallacy of power calculations for data analysis. American Statistician 55:19–24.

Holmes, R. T., and T. W. Sherry. 2001. Thirty-year bird population trends in an unfragmented temperate deciduous forest: importance of habitat change. Auk 118:589–609.

Hooper, R. G., J. C. Watson, and R. E. F. Escano. 1990. Hurricane Hugo's initial effects on red-cockaded woodpeckers in the Francis Marion National Forest. Transactions of the North American Wildlife Resources Conference 55:220–224.

Houghton, J. T., editor. 2001. Climate change 2001: the scientific basis. Contribution of working group I to the third assessment report of the intergovernmental panel on climate change. Cambridge University Press, Cambridge, U.K.

Howard, H. E. 1920. Territory in bird life. Murray, London.

Hudson, P. J., A. Rizzoli, B. T. Grenfell, H. Heesterbeck, and A. P. Dobson, editors. 2002. The ecology of wildlife diseases. Oxford University Press, Oxford.

Hughes, J., K. Goudkamp, D. Hurwood, M. Hancock, and S. Bunn. 2003. Translocation causes extinction of a local population of the freshwater shrimp *Paratya australiensis*. Conservation Biology 17:1007–1012.

Ingvarsson, P. K., and M. C. Whitlock. 2000. Heterosis increases the effective migration rate. Proceedings of the Royal Society of London B 267:1321–1326.

Jamieson, I. G., M. S. Roy, and M. Lettink. 2003. Sex-specific consequences of recent inbreeding in an ancestrally inbred population of New Zealand takahe. Conservation Biology 17:708–716.

Jansen, D., and T. Logan. 2002. Improving prospects for the Florida panther. United States Fish and Wildlife Service Endangered Species Bulletin 27:16–17.

Jenkins, P. F. 1978. Cultural transmission of song patterns and dialect development in a free-living bird population. Animal Behaviour 26:50–78.

Johnson, D. H. 1999. The insignificance of significance testing. Journal of Wildlife Management 63:763–772.

Johnston, R. F. 1956a. Population structure in salt marsh song sparrows. Part I: environment and annual cycle. Condor 58:24–44.

Johnston, R. F. 1956b. Population structure in salt marsh song sparrows. Part II: density, age structure and maintenance. Condor 58:254–272.

Jones, C. G., W. Heck, R. E. Lewis, Y. Mungroo, G. Slade, and T. Cade. 1995. The restoration of the Mauritius Kestrel *Falco punctatus* population. Ibis 137:S173–S180.

Joron, M., and P. M. Brakefield. 2003. Captivity masks inbreeding effects on male mating success in butterflies. Nature 424:191–194.

Kalinowski, S. T., and P. W. Hedrick. 1999. Detecting inbreeding depression is difficult in captive endangered species. Animal Conservation 2:131–136.

Kattan, G. H. 1997. Shiny cowbirds follow the "shotgun" strategy of brood parasitism. Animal Behaviour 53:647–654.

Keller, L. F. 1998. Inbreeding and its fitness effects in an insular population of song sparrows (*Melospiza melodia*). Evolution 52:240–250.

Keller, L. F., and P. Arcese. 1998. No evidence for inbreeding avoidance in a natural population of song sparrows (*Melospiza melodia*). American Naturalist 152:380–392.

Keller, L. F., P. Arcese, J. N. M. Smith, W. M. Hochachka, and S. C.

Stearns. 1994. Selection against inbred song sparrows during a natural population bottleneck. Nature 372:356–357.

Keller, L. F., P. R. Grant, B. R. Grant, and K. Petren. 2002. Environmental conditions affect the magnitude of inbreeding depression in survival of Darwin's finches. Evolution 56:1229–1239.

Keller, L. F., K. J. Jeffery, P. Arcese, M. A. Beaumont, W. M. Hochachka, J. N. M. Smith, and M. W. Bruford. 2001. Immigration and the ephemerality of a natural population bottleneck: evidence from molecular markers. Proceedings of the Royal Society of London B 268:1387–1394.

Keller, L. F., and D. M. Waller. 2002. Inbreeding effects in wild populations. Trends in Ecology and Evolution 17:230–241.

Kempenaers, B. 1995. Polygyny in the blue tit: intra- and inter-sexual conflicts. Animal Behaviour 49:1047–1064.

Kempenaers, B., G. R. Verheyen, and A. A. Dhondt. 1995. Mate guarding and copulation behaviour in monogamous and polygynous blue tits: do males follow a best-of-a-bad-job strategy? Behavioral Ecology and Sociobiology 36:33–42.

Kikkawa, J. 1980. Winter survival in relation to dominance classes among silvereyes *Zosterops lateralis chlorocephala* of Heron Island, Great Barrier Reef. Ibis 122:437–446.

Knapton, R. W., and J. R. Krebs. 1974. Settlement patterns, territory size, and breeding density in the song sparrow (*Melospiza melodia*). Canadian Journal of Zoology 52:1413–1420.

Knapton, R. W., and J. R. Krebs. 1976. Dominance hierarchies in winter song sparrows. Condor 78:567–569.

Knight, T. 1799. An account of some experiments on the fecundation of vegetables. Philisophical Transactions of the Royal Society of London 89:195–204.

Koenig, W. D., J. Haydock, and M. T. Stanback. 1998. Reproductive roles in the cooperatively breeding acorn woodpecker: incest avoidance versus reproductive competition. American Naturalist 151:243–255.

Koenig, W. D., D. van Vuren, and P. N. Hooge. 1996. Detectability, philopatry, and the distribution of dispersal distances in vertebrates. Trends in Ecology and Evolution 11:514–517.

Kölreuter, J. G. 1766. Vorläufige Nachricht von einigen das Geschlecht der Pflanzen betreffenden Versuchen und Beobachtungen. Leipzig.

Komdeur, J. 1996. Breeding of the Seychelles magpie robin *Copsychus sechellarum* and implications for its conservation. Ibis 138:485–498.

Komdeur, J. 1997. Inter-island transfers and population dynamics of Seychelles warblers *Acrocephalus sechellensis*. Bird Conservation International 7:7–26.

Komdeur, J., T. Piersma, K. Kraaijeveld, F. Kraaijeveld-Smit, and D. S. Richardson. 2004. Why Seychelles Warblers fail to recolonize nearby islands: unwilling or unable to fly there? Ibis 146:298–302.

Krebs, C. J. 2001. Ecology: the experimental analysis of distribution and abundance, 5th ed. Benjamin Cummings, San Francisco.

Krebs, C. J. 2002. Two complementary paradigms for analysing population dynamics. Philosophical Transactions of the Royal Society of London B 357:1211–1219.

Krebs, J. R. 1971. Territory and breeding density in the great tit, *Parus major* L. Ecology 52:2–22.

Krebs, J. R., and N. B. Davies. 1987. An introduction to behavioural ecology, 2nd ed. Blackwell Scientific Publications, Oxford.

Kroodsma, D. E. 1996. Ecology of passerine song development. Pages 3–19 *in* D. E. Kroodsma and E. H. Miller, editors. Ecology and evolution of acoustic communication in birds. Cornell University Press, Ithaca, New York.

Kruuk, L. E. B., B. C. Sheldon, and J. Merilä. 2002. Severe inbreeding depression in collared flycatchers (*Ficedula albicollis*). Proceedings of the Royal Society of London B 269:1581–1589.

Kuussaari, M., M. Nieminen, and I. Hanski. 1996. An experimental study of migration in the Glanville fritillary butterfly *Melitaea cinxia*. Journal of Animal Ecology 65:791–801.

Kuussaari, M., I. Saccheri, M. Camara, and I. Hanski. 1998. Allee effect and population dynamics in the Glanville fritillary butterfly. Oikos 82:384–392.

Lack, D. 1954. The natural regulation of animal numbers. Clarendon Press, Oxford.

Lack, D. 1966. Population study of birds. Clarendon Press, Oxford.

Lambin, X., J. Aars, and S. B. Piertney. 2001. Dispersal, intraspecific competition, kin competition and kin facilitation: a review of the empirical evidence. Pages 110–122 *in* J. Clobert, E. Danchin, A. A. Dhondt, and J. D. Nichols, editors. Dispersal. Oxford University Press, Oxford.

Lande, R. 1988a. Genetics and demography in biological conservation. Science 241:1455–1460.

Lande, R. 1988b. Demographic models of the northern spotted owl *Strix occidentalis caurina*. Oecologia 75:601–607.

Lande, R. 1998. Anthropogenic, ecological and genetic factors in extinction and conservation. Researches on Population Ecology 40:259–269.

Lande, R., S. Engen, and B. E. Sæther. 2003. Stochastic population dynamics in ecology and conservation. Oxford University Press, Oxford.

Laurance, W. F., K. R. McDonald, and R. Speare. 1996. Epidemic disease and the catastrophic decline of Australian rain forest frogs. Conservation Biology 10:406–413.

Leader-Williams, N. 1988. Reindeer on South Georgia. Cambridge University Press, Cambridge.

Lebreton, J. D., K. P. Burnham, J. Clobert, and D. R. Anderson. 1992. Modeling survival and testing biological hypotheses using marked animals: a unified approach with case studies. Ecological Monographs 62:67–118.

Leigh, E. G., Jr. 1975. Population fluctuations, community stability, and environmental variability. Pages 51–73 *in* M. L. Cody and J. M. Diamond, editors. Ecology and evolution of communities. Belknap Press of Harvard University Press, Cambridge, Massachusetts.

Leigh, E. G., Jr. 1981. The average lifetime of a population in a fluctuating environment. Journal of Theoretical Biology 90:231–239.

Lenormand, T. 2002. Gene flow and the limits to natural selection. Trends in Ecology and Evolution 17:183–189.

Levins, R. 1966. The strategy of model building in population biology. American Scientist 54:421–431.

Levins, R. 1969. Some demographic and genetic consequences of environmental heterogeneity for biological control. Bulletin of the Entomological Society of America 15:237–240.

Liebhold, A., and J. Bascompte. 2003. The Allee effect, stochastic dynamics and the eradication of alien species. Ecology Letters 6:133–140.

Lindenmayer, D. B., T. W. Clark, R. C. Lacy, and V. C. Thomas. 1993. Population viability analysis as a tool in wildlife management: a review with reference to Australia. Environmental Management 17:745–758.

Link, W. A., E. G. Cooch, and E. Cam. 2002. Model-based estimation of individual fitness. Journal of Applied Statistics 29:207–224.

Lomnicki, A. 1980. Regulation of population density due to individual differences and patchy environment. Oikos 35:185–193.

Lorenzana, J. C., and S. G. Sealy. 1999. A meta-analysis of the impact of parasitism by the brown-headed cowbird on its hosts. Studies in Avian Biology 18:241–253.

Lowther, P. E. 1993. Brown-headed cowbird (*Molothrus ater*). Pages 1–24 *in* A. F. Poole and F. B. Gill, editors. The birds of North America. No. 47. Birds of North America Inc., Philadelphia, Pennsylvania.

Ludwig, D. 1999. Is it meaningful to estimate a probability of extinction? Ecology 80:298–310.

Ludwig, D., and C. Walters. 2002. Fitting population viability analysis into adaptive management. Pages 511–520 *in* S. R. Beissinger and D. R. McCullough, editors. Population viability analysis. University of Chicago Press, Chicago, Illinois.

Lynch, M., J. Connery, and R. Bürger. 1995. Mutation accumulation and the extinction of small populations. American Naturalist 146:489–518.

Lynch, M., and B. Walsh. 1998. Genetics and analysis of quantitative traits. Sinauer Associates, Sunderland, Massachusetts.

MacArthur, R. H., and E. O. Wilson. 1967. The theory of island biogeography. Princeton University Press, Princeton, New Jersey.

Madsen, T., R. Shine, M. Olsson, and H. Wittzell. 1999. Restoration of an inbred adder population. Nature 402:34–35.

Maehr, D. S., and R. C. Lacy. 2002. Avoiding the lurking pitfalls in Florida panther recovery. Wildlife Society Bulletin 30:971–978.

Magrath, R. D., and R. G. Heinsohn. 2000. Reproductive skew in birds: models, problems and prospects. Journal of Avian Biology 31:247–258.

Malthus, T. R. 1798. An essay on the principle of population. J. Johnson, London.

Marr, A. B., L. C. Dallaire, and L. F. Keller. in press. Pedigree errors bias estimates of inbreeding depression. Animal Conservation, in press.

Marr, A. B., L. F. Keller, and P. Arcese. 2002. Heterosis and outbreeding depression in descendants of natural immigrants to an inbred population of song sparrows (*Melospiza melodia*). Evolution 56:131–142.

Marra, P. P. 2000. The role of behavioral dominance in structuring patterns of habitat occupancy in a migrant bird during the non-breeding season. Behavioral Ecology 11:299–308.

Marshall, J. A. 1948. Ecologic races of the song sparrow in the San Francisco Bay region. I. Habitat and abundance. Condor 50:193–215.

Martin, T. E., and J. Clobert. 1996. Nest predation and avian life-history evolution in Europe versus North America: a possible role of humans? American Naturalist 147:1028–1046.

Massot, M., J. Clobert, T. Pilorge, J. Lecomte, and R. Barbault. 1992. Density dependence in the common lizard—demographic consequences of a density manipulation. Ecology 73:1742–1756.

May, R. M., and S. K. Robinson. 1985. Population dynamics of avian brood parasitism. American Naturalist 126:475–494.

McBride, R. 2000. Current panther distribution and habitat use a review of field notes fall 1999–winter 2000. Florida Fish and Wildlife Commission Contract No. 95128. Livestock Protection Company, Alpine, Texas.

McCallum, H., J. Kikkawa, and C. Catterall. 2000. Density dependence in an island population of silvereyes. Ecology Letters 3:95–100.

McCleery, R. H., and C. M. Perrins. 1991. The effects of predation on the numbers of great tits. Pages 129–147 *in* C. M. Perrins, J. D. Lebreton, and G. J. M. Hirons, editors. Bird population studies. Relevance to conservation and management. Oxford University Press, Oxford.

McDonald, D. B. 1993. Delayed plumage maturation and orderly queues for status: a manakin mannequin experiment. Ethology 94:31–45.

McGregor, P. K., J. R. Krebs, and C. M. Perrins. 1981. Song repertoires and lifetime reproductive success in the great tit (*Parus major*). American Naturalist 118:149–159.

McLaren, C. M., and S. G. Sealy. 2003. Factors influencing susceptibility of host nests to brood parasitism. Ethology Ecology and Evolution 15:343–353.

McLaren, C. M., B. E. Woolfenden, H. L. Gibbs, and S. G. Sealy. 2003. Genetic and temporal patterns of multiple parasitism by brown-headed cowbirds (*Molothrus ater*) on song sparrows (*Melospiza melodia*). Canadian Journal of Zoology 81:281–286.

McMaster, D. G., and S. G. Sealy. 1998. Short incubation periods of brown-headed cowbirds: how do cowbird eggs hatch before yellow warbler eggs? Condor 100:102–111.

Merkt, J. R. 1981. An experimental study of habitat selection by the deer mouse, *Peromyscus maniculatus*, on Mandarte Island, B.C. Canadian Journal of Zoology 59:589–597.

Metcalfe, N. B., and P. Monaghan. 2001. Compensation for a bad start: grow now, pay later? Trends in Ecology and Evolution 16:254–260.

Meyer, A. H., B. R. Schmidt, and K. Grossenbacher. 1998. Analysis of three amphibian populations with quarter-century long time-series. Proceedings of the Royal Society of London B 265:523–528.

Miller, B., K. Ralls, R. P. Reading, J. M. Scott, and J. Estes. 1999. Biological and technical considerations of carnivore translocation: a review. Animal Conservation 2:59–68.

Mills, L. S., and F. W. Allendorf. 1996. The one-migrant-per-generation rule in conservation and management. Conservation Biology 10:1509–1518.

Milner, J. M., D. A. Elston, and S. D. Albon. 1999. Estimating the contributions of population density and climatic fluctuations to interannual variation in survival of Soay sheep. Journal of Animal Ecology 68:1235–1247.

Milner-Gulland, E. J., J. M. Pemberton, S. Brotherstone, and S. D. Albon. 2000. Estimating variance components and heritabilities in the wild: a case study using the "animal model" approach. Journal of Evolutionary Biology 13:804–813.

Møller, A. P., and F. de Lope. 1999. Senescence in a short-lived migratory bird: age dependent morphology, migration, reproduction and parasitism. Journal of Animal Ecology 68:163–171.

Monello, R. J., D. L. Murray, and E. F. Cassirer. 2001. Ecological correlates of pneumonia epizootics in bighorn sheep herds. Canadian Journal of Zoology 79:1423–1432.

Morris, W. F., and D. F. Doak. 2002. Quantitative conservation biology. Theory and practice of population viability analysis. Sinauer Associates, Sunderland, Massachusetts.

Morrison, M. L., L. S. Hall, S. K. Robinson, S. I. Rothstein, D. C. Hahn, and T. D. Rich, editors. 1999. Research and management of the brown-headed cowbird in western landscapes. Studies in Avian Biology 18: 1–322.

Mougeot, F., S. M. Redpath, R. Moss, J. Matthiopoulos, and P. J. Hudson. 2003. Territorial behaviour and population dynamics in red grouse *Lagopus lagopus scoticus*. I. Population experiments. Journal of Animal Ecology 72:1073–1082.

Mulligan, J. A. 1966. Singing behavior and its development in the song sparrow *Melospiza melodia*. University of California Publications in Zoology 81:1–76.

Murphy, M. T. 1996. Survivorship, breeding dispersal and mate fidelity in eastern kingbirds. Condor 98:82–92.

Myers, J. H., D. Simberloff, A. M. Kuris, and J. R. Carey. 2000. Eradication revisited: dealing with exotic species. Trends in Ecology and Evolution 15:316–320.

Naef-Daenzer, B., F. Widmer, and M. Nuber. 2001. Differential post-fledging survival of great and coal tits in relation to their condition and fledging date. Journal of Animal Ecology 70:730–738.

Nakamura, T. K., and A. Cruz. 2000. The ecology of egg-puncture behavior by the shiny cowbird in southwestern Puerto Rico. Pages 178–186 *in* J. N. M. Smith, T. L. Cook, S. I. Rothstein, S. K. Robinson, and S. G. Sealy, editors. Ecology and management of cowbirds and their hosts. University of Texas Press, Austin, Texas.

Nelson, J. B. 1978. The gannett. T. and A. D. Poyser, Berkhamsted, U.K.

Newton, I., editor. 1989a. Lifetime reproduction in birds. Academic Press, London.

Newton, I. 1989b. Introduction. Pages 1–11 *in* I. Newton, editor. Lifetime reproduction in birds. Academic Press, London.

Newton, I. 1991. Population limitation in birds of prey: a comparative approach. Pages 542–565 *in* C. M. Perrins, J. D. Lebreton, and G. J. M. Hirons, editors. Bird population studies. Relevance to conservation and management. Oxford University Press, Oxford.

Newton, I. 1992. Experiments on the limitation of bird numbers by territorial behavior. Biological Reviews of the Cambridge Philosophical Society 67:129–173.

Newton, I. 1998. Population limitation in birds. Academic Press, San Diego.

Nice, M. M. 1937. Studies in the life history of song sparrows. I. Transactions of the Linnean Society of New York 4:1–273.

Nice, M. M. 1941. The role of territory in bird life. American Midland Naturalist 26:441–487.

Nice, M. M. 1943. Studies in the life history of song sparrows. II. Transactions of the Linnean Society of New York 6:1–328.

Nieminen, M., M. C. Singer, W. Fortelius, K. Schops, and I. Hanski. 2001. Experimental confirmation that inbreeding depression increases extinction risk in butterfly populations. American Naturalist 157:237–244.

Nol, E., and J. N. M. Smith. 1987. Effects of age and breeding experience on seasonal reproductive success in the song sparrow. Journal of Animal Ecology 56:301–313.

Nolan, V., Jr. 1978. The ecology and behavior of the prairie warbler, *Dendroica discolor*. Ornithological Monographs 26:1–595.

Nolan, V., Jr., and E. D. Ketterson. 1990. Timing of autumn migration and its relation to winter distribution in dark-eyed juncos. Ecology 71:1267–1278.

Nordby, J. C., S. E. Campbell, and M. D. Beecher. 1999. Ecological correlates of song learning in song sparrows. Behavioral Ecology 10:287–297.

Nowicki, S., S. Peters, and J. Podos. 1998. Song learning, early nutrition and sexual selection in songbirds. American Zoologist 38:179–190.

Nunney, L., and D. R. Elam. 1994. Estimating the effective population size of conserved populations. Conservation Biology 8:175–184.

O'Brien, S. J., and J. F. Evermann. 1988. Interactive influence of infectious disease and genetic diversity in natural populations. Trends in Ecology and Evolution 3:254–259.

O'Connor, K. D. 2003. Extra-pair mating and effective population size in the song sparrow (*Melospiza melodia*). M.Sc. Thesis. University of British Columbia, Vancouver.

O'Connor, K.D., A.B. Marr, P. Arcese, L.F. Keller, K.J. Jeffery, and M.W. Bruford. Extra-pair fertilization and effective population size in the song sparrow (*Melospiza melodia*). Journal of Avian Biology, in press.

Olesen, J. M., and A. Valido. 2003. Lizards as pollinators and seed dispersers: an island phenomenon. Trends in Ecology and Evolution 18:177–181.

Oppel, S., H. M. Schaefer, V. Schmidt, and B. Schröder. 2004. Cowbird parasitism of pale-headed brush-finch *Atapeltes pallidiceps*. Bird Conservation International 14:63–75.

Ortega, C. P. 1998. Cowbirds and other brood parasites. University of Arizona Press, Tucson, Arizona.

Parker, P. G., and T. A. Waite. 1997. Mating systems, effective population size, and conservation of natural populations. Pages 243–261 *in* J. R. Clemmons and R. Buchholz, editors. Behavioral approaches to conservation in the wild. Cambridge University Press, Cambridge.

Parmesan, C. 1996. Climate and species' range. Nature 382:765–766.

Partridge, L. 1989. Lifetime reproductive success and life-history evolution. Pages 421–440 *in* I. Newton, editor. Lifetime reproduction in birds. Academic Press, London.

Patten, M. A., J. T. Rotenberry, and M. Zuk. 2004. Habitat selection, acoustic adaptation, and the evolution of reproductive isolation. Evolution 58:2144–2155.

Payne, R. B. 1997. Avian brood parasitism. Pages 338–369 *in* D. H. Clayton and J. Moore, editors. Host-parasite evolution: general principles and avian models. Oxford University Press, Oxford.

Payne, R. B., and L. L. Payne. 1996. Dispersal, demography and the persistence of partnerships in the indigo bunting. Pages 305–320 *in* J. M. Black, editor. Partnerships in birds: the study of monogamy. Oxford University Press, Oxford.

Pease, C. M., and J. A. Grzybowski. 1995. Assessing the consequences of brood parasitism and nest predation on seasonal fecundity in passerine birds. Auk 112:343–363.

Perrins, C. M., and R. H. McCleery. 1994. Competition and egg weight in the great tit *Parus major*. Ibis 136:454–456.

Peterjohn, B. G., J. R. Sauer, and S. Schwarz. 2000. Temporal and geographic patterns in population trends of brown-headed cowbirds. Pages 21–34 *in* J. N. M. Smith, T. L. Cook, S. I. Rothstein, S. K. Robinson, and S. G. Sealy, editors. Ecology and management of cowbirds and their hosts. University of Texas Press, Austin, Texas.

Petren, K., and T. J. Case. 1996. An experimental demonstration of exploitation competition in an ongoing invasion. Ecology 77:118–132.

Petrinovich, L., T. Patterson, and L. F. Baptista. 1981. Song dialects as barriers to dispersal, a reevaluation. Evolution 35:180–188.

Pimm, S. L., and O. L. Bass, Jr. 2002. Rangewide risks to large populations: the Cape Sable sparrow case history. Pages 406–424 *in* S. R. Beissinger and D. R. McCullough, editors. Population viability analysis. Chicago University Press, Chicago, Illinois.

Pimm, S. L., H. L. Jones, and J. Diamond. 1988. On the risk of extinction. American Naturalist 132:757–785.

Piper, W. H. 1997. Social dominance in birds: early findings and new horizons. Current Ornithology 14:125–187.

Podos, J., S. Peters, T. Rudnicky, P. Marler, and S. Nowicki. 1992. The organization of song repertoires in song sparrows: themes and variations. Ethology 90:89–106.

Post, W., and J. S. Greenlaw. 1994. Seaside sparrow. Pages 1–28 *in* A. F. Poole and F. B. Gill, editors. The birds of North America: life histories for the 21st century. No. 127. American Ornithologist's Union, Philadelphia, Pennsylvania.

Postma, E., and A. J. van Noordwijk. 2005. Gene flow maintains a large genetic difference in clutch size at a small spatial scale. Nature 433:65–68.

Pusey, A., and M. Wolf. 1996. Inbreeding avoidance in animals. Trends in Ecology and Evolution 11:201–206.

Puurtinen, M., K. E. Knott, S. Suonpaa, T. van Ooik, and V. Kaitala. 2004. Genetic variability and drift load in populations of an aquatic snail. Evolution 58:749–756.

Ralls, K., J. D. Ballou, and A. Templeton. 1988. Estimates of lethal equivalents and the cost of inbreeding in mammals. Conservation Biology 2:185–193.

Ralls, K., S. R. Beissinger, and J. F. Cochrane. 2002. Guidelines for using population viability analysis in endangered-species management. Pages 521–550 *in* S. R. Beissinger and D. R. McCullough, editors. Population viability analysis. Chicago University Press, Chicago, Illinois.

Ralls, K., K. Brugger, and J. Ballou. 1979. Inbreeding and juvenile mortality in small populations of ungulates. Science 206:1101–1103.

Ralls, K., P. H. Harvey, and A. M. Lyles. 1986. Inbreeding in natural populations of birds and mammals. Pages 35–56 *in* M. E. Soule, editor. Conservation biology: the science of scarcity and diversity. Sinauer Associates Inc., Sunderland, Massachusetts.

Rannala, B., and J. L. Mountain. 1997. Detecting immigration by using multilocus genotypes. Proceedings of the National Academy of Sciences of the United States of America 94:9197–9201.

Rasa, O. A. E. 1989. The costs and effectiveness of vigilance behavior in the dwarf mongoose—implications for fitness and optimal group size. Ethology, Ecology, and Evolution 1:265–282.

Reddingius, J., and P. J. Den Boer. 1970. Simulation experiments illustrating stabilization of animal numbers by spreading of risk. Oecologia 5:240–284.

Reed, D. H., E. H. Lowe, D. A. Briscoe, and R. Frankham. 2003. Inbreeding and extinction: Effects of rate of inbreeding. Conservation Genetics 4:405–410.

Reed, J. M., L. S. Mills, J. B. Dunning, Jr., E. S. Menges, K. S. McKelvey, R. Frye, et al. 2002. Emerging issues in population viability analysis. Conservation Biology 16:7–19.

Reid, J. M., P. Arcese, A. L. E. V. Cassidy, S. M. Hiebert, J. N. M. Smith, P. K. Stoddard, et al. 2004. Song repertoire size predicts initial mating success in male song sparrows (*Melospiza melodia*). Animal Behaviour 68:1055–1063.

Reid, J. M., P. Arcese, A. L. E. V. Cassidy, S. M. Hiebert, J. N. M. Smith, P. K. Stoddard, et al. 2005a. Fitness correlates of repertoire size in free-living song sparrows (*Melospiza melodia*). American Naturalist 165:299–310.

Reid, J. M., P. Arcese, A. L. E. V. Cassidy, A. B. Marr, J. N. M. Smith, and L. F. Keller. 2005b. Hamilton and Zuk meet heterozygosity? Song repertoire size signals inbreeding and immunity in song sparrows (*Melospiza melodia*). Proceedings of the Royal Society of London B 282:481–487.

Reid, J. M., P. Arcese, and L. F. Keller. 2003a. Inbreeding depresses immune response in song sparrows (*Melospiza melodia*): direct and intergenerational effects. Proceedings of the Royal Society of London B 270:2151–2157.

Reid, J. M., E. M. Bignal, S. Bignal, D. I. McCracken, and P. Monaghan. 2003b. Age-specific reproductive performance in red-billed choughs (*Pyrrhocorax pyrrhocorax*): patterns and processes in a natural population. Journal of Animal Ecology 72:765–776.

Richards, C. M. 2000. Inbreeding depression and genetic rescue in a plant metapopulation. American Naturalist 155:383–394.

Richison, G. 1983. The function of singing in female black-headed grosbeaks (*Pheucticus melanocephalus*). Auk 100:105–116.

Richter-Dyn, N., and N. S. Goel. 1972. On the extinction of a colonizing species. Theoretical Population Biology 3:406–433.

Ricklefs, R. E. 2000. Density dependence, evolutionary optimization, and the diversification of avian life histories. Condor 102:9–22.

Ricklefs, R. E., and M. Wikelski. 2002. The physiology/life-history nexus. Trends in Ecology and Evolution 17:462–468.

Rising, J. D. 1996. A guide to the identification and natural history of the sparrows of the United States and Canada. Academic Press, San Diego, California.

Robbins, C. S., J. R. Sauer, R. S. Greenberg, and S. Droege. 1989. Population declines in North American birds that migrate to the Neotropics. Proceedings of the National Academy of Sciences of the United States of America 86:7658–7662.

Robertson, B. C., S. M. Degnan, J. Kikkawa, and C. C. Moritz. 2001. Genetic monogamy in the absence of paternity guards: the Capricorn silvereye, *Zosterops lateralis chlorocephalus*, on Heron Island. Behavioral Ecology 12:666–673.

Robertson, R. J., and W. B. Rendell. 2001. A long-term study of reproductive performance in tree swallows: the influence of age and senescence on output. Journal of Animal Ecology 70:1014–1031.

Robinson, S. K., J. P. Hoover, and J. R. Herkert. 2000. Cowbird parasitism in a fragmented landscape: effects of tract size, habitat, and abundance of cowbirds and hosts. Pages 323–332 *in* J. N. M. Smith, T. L. Cook, S. I. Rothstein, S. K. Robinson, and S. G. Sealy, editors. Ecology and management of cowbirds and their hosts. University of Texas Press, Austin, Texas.

Rogers, C. M., J. N. M. Smith, W. M. Hochachka, A. L. E. V. Cassidy, M. J. Taitt, P. Arcese, and D. Schluter. 1991. Spatial variation in winter survival of song sparrows *Melospiza melodia*. Ornis Scandinavica 22:387–395.

Rogers, C. M., M. J. Taitt, J. N. M. Smith, and G. Jongejan. 1997. Nest predation and cowbird parasitism create a demographic sink in wetland-breeding song sparrows. Condor 99:622–633.

Rohner, C. 2004. Night moves. Living Bird 23:24–29.

Rosenzweig, M. L. 2003. Win-win ecology: how the earth's species can survive in the midst of human enterprise. Oxford University Press, New York.

Rothstein, S. I. 1975. An experimental and teleonomic investigation of avian brood parasitism. Condor 77:250–271.

Rothstein, S. I. 1994. The cowbird's invasion of the far west: history, causes and consequences experienced by host species. Studies in Avian Biology 15:301–315.

Rothstein, S. I., and T. L. Cook. 2000. Cowbird management, host population limitation, and efforts to save endangered species. Pages 323–332 *in* J. N. M. Smith, T. L. Cook, S. I. Rothstein, S. K. Robinson, and S. G. Sealy, editors. Ecology and management of cowbirds and their hosts. University of Texas Press, Austin, Texas.

Rothstein, S. I., and S. K. Robinson, editors. 1998a. Parasitic birds and their hosts, studies in coevolution. Oxford University Press, Oxford and New York.

Rothstein, S. I., and S. K. Robinson. 1998b. The evolution and ecology of avian brood parasitism, an overview. Pages 3–56 *in* S. I. Rothstein and S. K. Robinson, editors. Parasitic birds and their hosts, studies in coevolution. Oxford University Press, Oxford and New York.

Rowe, G., and T. J. C. Beebee. 2003. Population on the verge of a mutational meltdown? Fitness costs of genetic load for an amphibian in the wild. Evolution 57:177–181.

Ryan, K. K., and R. C. Lacy. 2003. Monogamous male mice bias behaviour towards females according to very small differences in kinship. Animal Behaviour 65:379–384.

Saccheri, I., M. Kuussaari, M. Kankare, P. Vikman, W. Fortelius, and I. Hanski. 1998. Inbreeding and extinction in a butterfly metapopulation. Nature 392:491–494.

Saccheri, I. J., and P. M. Brakefield. 2002. Rapid spread of immigrant genomes into inbred populations. Proceedings of the Royal Society of London B 269:1073–1078.

Sæther, B. E. 1990. Age-specific variation in reproductive performance in birds. Current Ornithology 7:251–283.

Sæther, B. E., S. Engen, R. Lande, P. Arcese, and J. N. M. Smith. 2000a. Estimating the time to extinction in an island population of song sparrows. Proceedings of the Royal Society of London B 267:621–626.

Sæther, B. E., S. Engen, and E. Matthysen. 2002. Demographic characteristics and population dynamical patterns of solitary birds. Science 295:2070–2073.

Sæther, B. E., J. Tufto, S. Engen, K. Jerstad, O. W. Røstad, and J. E. Skåtan. 2000b. Population dynamical consequences of climate change for a small temperate songbird. Science 287:854–856.

Sanders, N. J., N. J. Gotelli, N. E. Heller, and D. M. Gordon. 2003. Community disassembly by an invasive species. Proceedings of the National Academy of Sciences of the United States of America 100:2474–2477.

Sauer, J. R., J. E. Hines, and J. Fallon. 2005. The North American Breeding Bird Survey, Results and Analysis 1966–2004. Version 2005.2. USGS Patuxent Wildlife Research Center, Laurel, Maryland

Saunders, C. A., P. Arcese, and K. D. O'Connor. 2003. Nest site characteristics in the song sparrow and parasitism by brown-headed cowbirds. Wilson Bulletin 115–28.

Savidge, J. A. 1987. Extinction of an island forest avifauna by an introduced snake. Ecology 68:660–668.

Schluter, D., and L. Gustafsson. 1993. Maternal inheritance of condition and clutch size in the collared flycatcher. Evolution 47:658–667.

Schluter, D., and J. N. M. Smith. 1986. Genetic and phenotypic correlations in a natural population of song sparrows. Biological Journal of the Linnean Society 29:23–36.

Schreiber, E. A., and R. W. Schreiber. 1989. Insights into seabird biology from a global natural experiment. National Geographic Research 5:64–81.

Schroeder, M. A., J. R. Young, and C. E. Braun. 1999. Sage grouse (*Centrocercus urophasianus*). Pages 1–28 *in* A. F. Poole and F. B. Gill, editors. The birds of North America: life histories for the 21st century. No. 425. American Ornithologist's Union, Philadelphia, Pennsylvania.

Scott, D. M., and C. D. Ankney. 1983. The laying cycle of brown-headed cowbirds: passerine chickens? Auk 100:583–592.

Scott, D. M., P. J. Weatherhead, and C. D. Ankney. 1992. Egg-eating by female brown-headed cowbirds. Condor 94:579–584.

Sealy, S. G., D. G. McMaster, S. A. Gill, and D. L. Neudorf. 2000. Yellow warbler nest attentiveness before sunrise: antiparasite strategy or onset of incubation? Pages 169–177 *in* J. N. M. Smith, T. L. Cook, S. I. Rothstein, S. K. Robinson, and S. G. Sealy, editors. Ecology and management of cowbirds and their hosts. University of Texas Press, Austin, Texas.

Searcy, W. A. 1984. Song repertoire size and female preferences in song sparrows. Behavioral Ecology and Sociobiology 14:281–286.

Searcy, W. A., and P. Marler. 1981. A test for responsiveness to song structure and programming in female sparrows. Science 213:926–928.

Searcy, W. A., P. D. McArthur, and K. Yasukawa. 1985. Song repertoire size and male quality in song sparrows. Condor 87:222–228.

Searcy, W. A., S. Nowicki, M. Hughes, and S. Peters. 2002. Geographic song discrimination in relation to dispersal distances in the song sparrow. American Naturalist 159:221–230.

Searcy, W. A., S. Nowicki, and S. Peters. 1999. Song types as fundamental units in vocal repertoires. Animal Behaviour 58:37–44.

Searcy, W. A., and K. Yasukawa. 1995. Polygyny and sexual selection in the red-winged blackbird. Princeton University Press, Princeton, New Jersey.

Searcy, W. A., and K. Yasukawa. 1996. Song and female choice. Pages 454–473 *in* D. E. Kroodsma and E. H. Miller, editors. Ecology and evolution of acoustic communication in birds. Cornell University Press, Ithaca, New York.

Sedgwick, J. A., and W. M. Iko. 1999. Costs of brown-headed cowbird parasitism to willow flycatchers. Studies in Avian Biology 18:167–181.

Shaffer, M. L. 1981. Minimum population sizes for species conservation. BioScience 31:131–134.

Shaffer, M. L. 1983. Determining minimum population sizes for the grizzly bear. International Conference on Bear Research and Management 5:133–139.

Shaffer, M. L., and F. B. Samson. 1985. Population size and extinction. A note on determining critical population sizes. American Naturalist 125:144–152.

Shields, W. M. 1993. The natural and unnatural history of inbreeding and outbreeding. Pages 143–169 *in* N. W. Thornhill, editor. The natural history of inbreeding and outbreeding: theoretical and empirical perspectives. University of Chicago Press, Chicago, Illinois.

Shull, G. H. 1914. Duplicated genes for capsule form in *Bursa bursapastoris*. Zeitschrift für induktive Abstammungs-und Vererbungslehre 12:97–149.

Shuster, S. M., and M. J. Wade. 2003. Mating systems and strategies. Princeton University Press, Princeton, New Jersey.

Sibley, C. G., and J. Ahlquist. 1990. Phylogeny and classification of birds: a study in molecular evolution. Yale University Press, New Haven, Connecticut.

Sillett, T. S., and R. T. Holmes. 2002. Variation in survivorship of a migratory songbird throughout its annual cycle. Journal of Animal Ecology 71:296–308.

Sillett, T. S., R. T. Holmes, and T. W. Sherry. 2000. Impacts of a global climate cycle on population dynamics of a migratory songbird. Science 288:2040–2042.

Simberloff, D. 1988. The contribution of population and community biology to conservation science. Annual Review of Ecology and Systematics 19:473–511.

Sinclair, A. R. E. 1989. Population regulation in animals. Pages 197–241 *in* J. M. Cherrett, editor. Ecological concepts. Blackwell Scientific Publications, Oxford.

Sinclair, A. R. E., R. P. Pech, C. R. Dickman, D. Hik, P. Mahon, and A. E. Newsome. 1998. Predicting effects of predation on conservation of endangered prey. Conservation Biology 12:564–575.

Skutch, A. F. 1954. Life histories of Central American birds. Vol. 1. Pacific Coast Avifauna 31. Cooper Ornithological Society, Berkeley, California.

Skutch, A. F. 1960. Life histories of Central American birds. Vol. 2. Pacific Coast Avifauna 34. Cooper Ornithological Society, Berkeley, California.

Smith, J. A., K. Wilson, J. G. Pilkington, and J. M. Pemberton. 1999. Heritable variation in resistance to gastrointestinal nematodes in an unmanaged mammal population. Proceedings of the Royal Society of London B 266:1283–1290.

Smith, J. N. M. 1978. Division of labor by song sparrows feeding fledged young. Canadian Journal of Zoology 56:187–191.

Smith, J. N. M. 1981a. Cowbird parasitism, host fitness, and age of the host female in an island song sparrow population. Condor 83:152–161.

Smith, J. N. M. 1981b. Does high fecundity reduce survival in song sparrows? Evolution 35:1142–1148.

Smith, J. N. M. 1982. Song sparrow pair raise four broods in one year. Wilson Bulletin 94:585.

Smith, J. N. M. 1988. Determinants of lifetime reproductive success in the song sparrow. Pages 154–172 *in* T. H. Clutton-Brock, editor. Reproductive success. Chicago University Press, Chicago, Illinois.

Smith, J. N. M. 1994. Cowbirds: conservation villains or convenient scapegoats? Birding 26:257–259.

Smith, J. N. M. 1999. The basis for cowbird management: host selection, impacts on hosts, and criteria for taking management action. Studies in Avian Biology 18:104–108.

Smith, J. N. M., and P. Arcese. 1989. How fit are floaters—consequences of alternative territorial behaviours in a non-migratory sparrow. American Naturalist 133:830–845.

Smith, J. N. M., and P. Arcese. 1994. Brown-headed cowbirds and an island population of song sparrows: a 16-year study. Condor 96:916–934.

Smith, J. N. M., P. Arcese, and I. G. McLean. 1984. Age, experience, and enemy recognition by wild song sparrows. Behavioral Ecology and Sociobiology 14:101–106.

Smith, J. N. M., P. Arcese, and D. Schluter. 1986. Song sparrows grow and shrink with age. Auk 103:210–212.

Smith, J. N. M., T. L. Cook, S. I. Rothstein, S. K. Robinson, and S. G. Sealy, editors, 2000. Ecology and management of cowbirds and their hosts. University of Texas Press, Austin, Texas.

Smith, J. N. M., and J. R. Merkt. 1980. Development and stability of single-parent family units in the song sparrow. Canadian Journal of Zoology 58:1869–1875.

Smith, J. N. M., R. D. Montgomerie, M. J. Taitt, and Y. Yom-Tov. 1980. A winter feeding experiment on an island song sparrow population. Oecologia 47:164–170.

Smith, J. N. M., and I. H. Myers-Smith. 1998. Spatial variation in parasitism of song sparrows by brown-headed cowbirds. Pages 296–312 *in* S. I. Rothstein and S. K. Robinson, editors. Brood parasites and their hosts. Oxford University Press, Oxford.

Smith, J. N. M., and D. A. Roff. 1980. Temporal spacing of broods, brood size, and parental care in song sparrows (*Melospiza Melodia*). Canadian Journal of Zoology 58:1007–1015.

Smith, J. N. M., and S. I. Rothstein. 2000. Brown-headed cowbirds as model systems for the study of behavior, ecology, evolution and conservation biology. Pages 87–99 *in* J. N. M. Smith, T. L. Cook, S. I. Rothstein, S. K. Robinson, and S. G. Sealy, editors. Ecology and management of cowbirds and their hosts. University of Texas Press, Austin, Texas.

Smith, J. N. M., M. J. Taitt, C. M. Rogers, P. Arcese, L. F. Keller, A. L. E. V. Cassidy, and W. M. Hochachka. 1996. A metapopulation approach to the population biology of the song sparrow *Melospiza melodia*. Ibis 138:S120–S128.

Smith, J. N. M., M. J. Taitt, and L. Zanette. 2002. Removing brown-headed cowbirds increases seasonal fecundity and population growth in song sparrows. Ecology 83:3037–3047.

Smith, J. N. M., M. J. Taitt, L. Zanette, and I. H. Myers-Smith. 2003. How do brown-headed cowbirds (*Molothrus ater*) cause nest failures in song sparrows (*Melospiza melodia*)? A removal experiment. Auk 120:772–783.

Smith, J. N. M., Y. Yom-Tov, and R. Moses. 1982. Polygyny, male parental care, and sex ratio in song sparrows: an experimental study. Auk 99:555–564.

Smith, S. M. 1978. The "underworld" of a territorial sparrow: adaptive strategy for floaters. American Naturalist 112:571–582.

Smith, S. M. 1991. The black-capped chickadee. Cornell University Press, Ithaca, New York.

Soulé, M. E. 1980. Thresholds for survival: maintaining fitness and evolutionary potential. Pages 151–170 *in* M. E. Soulé and B. A. Wilcox, editors. Conservation biology: an evolutionary-ecological perspective. Sinauer Associates, Sunderland, Massachusetts.

Soulé, M. E., and B. A. Wilcox, editors. 1980. Conservation biology, an evolutionary-ecological perspective. Sinauer Associates, Sunderland, Massachusetts.

Spielman, D., B. W. Brook, and R. Frankham. 2004. Most species are not driven to extinction before genetic factors impact them. Proceedings of the National Academy of Sciences of the United States of America 101:15261–15264.

Stake, M. M., and D. A. Cimprich. 2003. Using video to monitor predation at black-capped vireo nests. Condor 105:348–357.

Steadman, D. W. 1995. Prehistoric extinction of pacific island birds: biodiversity meets zooarchaeology. Science 267:1123–1131.

Stearns, B. P., and S. C. Stearns. 1999. Watching from the edge of extinction. Yale University Press, New Haven, Connecticut.

Stephens, P. A., and W. J. Sutherland. 1999. Consequences of the Allee effect for behaviour, ecology and conservation. Trends in Ecology and Evolution 14:401–405.

Stewart, B. S., P. K. Yochem, H. R. Huber, R. L. De Long, R. J. Jameson, W. J. Sydeman, et al. 1994. History and present status of the northern elephant seal population. Pages 29–48 *in* B. J. Le Boeuf and R. M. Laws, editors. Elephant seals: population ecology, behavior, and physiology. University of California Press, Berkeley, California.

Stewart, R. E., and J. W. Aldrich. 1951. Removal and repopulation of breeding birds in a spruce-fir forest community. Auk 68:471–482.

Stirling, I., N. J. Lunn, J. Iacozza, C. Elliott, and M. Obbard. 2004. Polar bear distribution and abundance on the Southwestern Hudson Bay Coast during open water season, in relation to population trends and annual ice patterns. Arctic 57:15–26.

Stockwell, C. A., A. P. Hendry, and M. T. Kinnison. 2003. Contemporary evolution meets conservation biology. Trends in Ecology and Evolution 18:94–101.

Storfer, A. 1999. Gene flow and endangered species translocations: a topic revisited. Biological Conservation 87:173–180.

Strausberger, B. M., and M. V. Ashley. 2003. Breeding biology of brood parasitic brown-headed cowbirds (*Molothrus ater*) characterized by parent-offspring and sibling-group reconstruction. Auk 120:433–445.

Strausberger, B. M., and D. E. Burhans. 2001. Nest desertion by field sparrows and its possible influence on the evolution of cowbird behavior. Auk 118:770–776.

Strebel, D. E. 1985. Environmental fluctuations and extinction—single species. Theoretical Population Biology 27:1–26.

Stutchbury, B. J. M. 1997. Effects of female cowbird removal on reproductive success of hooded warblers. Wilson Bulletin 109:74–81.

Sures, B., and K. Knopf. 2004. Parasites as a threat to freshwater eels? Science 304:208–209.

Takasu, F., K. Kawasaki, H. Nakamura, J. E. Cohen, and N. Shigesada. 1993. Modeling the population dynamics of a cuckoo-host association and the evolution of host defenses. American Naturalist 142:819–839.

Tallmon, D. A., G. Luikart, and R. S. Waples. 2004. The alluring simplicity and complex reality of genetic rescue. Trends in Ecology and Evolution 19:489–496.

Telfer, S., S. B. Piertney, J. F. Dallas, W. A. Stewart, F. Marshall, J. L. Gow, and X. Lambin. 2003. Parentage assignment detects frequent and large-scale dispersal in water voles. Molecular Ecology 12:1939–1949.

Templeton, A. R. 1986. Coadaptation and outbreeding depression. Pages 105–116 *in* M. E. Soulé, editor. Conservation biology: the science of scarcity and diversity. Sinauer Associates, Sunderland, Massachusetts.

Terborgh, J. 1989. Where have all the birds gone. Princeton University Press, Princeton, New Jersey.

Thomas, C. D., A. Cameron, R. E. Green, M. Bakkenes, L. J. Beaumont, Y. C. Collingham, et al. 2004. Extinction risk from climate change. Nature 427:145–148.

Thompson, F. R. I., T. M. Donovan, R. M. Degraaf, J. Faaborg, and S. K. Robinson. 2002. A multi-scale perspective of the effects of forest fragmentation on birds in eastern forests. Studies in Avian Biology 25:8–19.

Thornhill, N. W. 1993. The natural history of inbreeding and outbreeding: theoretical and empirical perspectives. University of Chicago Press, Chicago, Illinois.

Tompa, F. S. 1963. Factors determining the numbers of song sparrows on Mandarte Island, B.C. Ph.D. thesis. University of British Columbia, Vancouver.

Tompa, F. S. 1964. Factors determining the numbers of song sparrows, *Melospiza melodia* (Wilson), on Mandarte island, B.C., Canada. Acta Zoologica Fennica 109:1–73.

Tompa, F. S. 1971. Catastrophic mortality and its population consequences. Auk 88:753–759.

Trine, C. L., W. D. Robinson, and S. K. Robinson. 1998. Consequences of brown-headed cowbird parasitism for host population dynamics. Pages

273–295 *in* S. I. Rothstein and S. K. Robinson, editors. Parasitic birds and their hosts. Studies in coevolution. Oxford University Press, Oxford.

Trine, C. L., W. D. Robinson, and S. K. Robinson. 2000. Effects of multiple parasitism on cowbird and wood thrush nesting success. Pages 135–144 *in* J. N. M. Smith, T. L. Cook, S. I. Rothstein, S. K. Robinson, and S. G. Sealy, editors. Ecology and management of cowbirds and their hosts. University of Texas Press, Austin, Texas.

Tufto, J., B. E. Sæther, S. Engen, P. Arcese, K. Jerstad, O. W. Røstad, and J. N. M. Smith. 2000. Bayesian meta-analysis of demographic parameters in three small, temperate passerines. Oikos 88:273–281.

van Balen, J. H. 1980. Population fluctuations of the great tit and feeding conditions in winter. Ardea 68:143–164.

van Balen, J. H., and R. P. J. Potting. 1990. Comparative reproductive biology of four blue tit populations in the Netherlands. Pages 91–102 *in* J. Blondel, A. Gosler, J. D. Lebreton, and R. H. McCleery, editors. Population Biology of Passerine Birds. NATO Advanced Science Institutes Series No. 24. Springer-Verlag, Berlin.

van Riper, C., S. G. van Riper, M. L. Goff, and M. Laird. 1986. The epizootiology and ecological significance of malaria in Hawaiian land birds. Ecological Monographs 56:327–344.

Verbeek, N. A. M. 1982. Egg predation by northwestern crows—its association with human and bald eagle (*Halilaetus leucocephalus*) activity. Auk 99:347–352.

Verhulst, S., and H. M. van Eck. 1996. Gene flow and immigration rate in an island population of great tits. Journal of Evolutionary Biology 9:771–782.

Verner, J., and M. F. Willson. 1969. Mating systems, sexual dimorphism and the role of male North American passerine birds in the nesting cycle. Ornithological Monographs 9:1–76.

Vilà, C., A. Sundqvist, Ø. Flagstad, J. Seddon, S. Björnerfeldt, I. Kojola, et al. 2003. Rescue of a severely bottlenecked wolf (*Canis lupus*) population by a single immigrant. Proceedings of the Royal Society of London B 270:91–97.

Waite, T. A., and P. G. Parker. 1997. Extrapair paternity and the effective size of socially monogamous populations. Evolution 51:620–621.

Waller, D. M., J. Dole, and A. Bersch. in press. Does stress increasethe expression of inbreeding depression? Experiments with *Brassica rapa*. Evolution.

Walters, C. 1986. Adaptive management of renewable resources. Macmillan, New York.

Walters, C. 1997. Challenges in adaptive management of riparian and coastal ecosystems. Conservation Ecology [online] 1(2). Available at www.ecologyandsociety.org/vol1/iss2/art1/.

Watson, A. 1985. Social class, socially induced loss, recruitment and breeding of red grouse *Lagopus lagopus scoticus*. Oecologia 67:493–498.

Watson, A., and R. Moss. 1970. Dominance, spacing behaviour and aggression in relation to population limitation in vertebrates. Pages 167–218 *in* A. Watson, editor. Animal populations in relation to their food resources. Blackwell Scientific Publications, Oxford.

Weatherhead, P. J., and K. A. Boak. 1986. Site infidelity in song sparrows. Animal Behaviour 34:1299–1310.

Weidenfeld, D. A. 2000. Cowbird population changes and their relationship to some changes in host species. Pages 35–46 *in* J. N. M. Smith, T. L. Cook, S. I. Rothstein, S. K. Robinson, and S. G. Sealy, editors. Ecology and management of cowbirds and their hosts. University of Texas Press, Austin, Texas.

Westemeier, R. L., J. D. Brawn, S. A. Simpson, T. L. Esker, R. W. Jansen, J. W. Walk, et al. 1998. Tracking the long-term decline and recovery of an isolated population. Science 282:1695–1698.

Westneat, D. F. 1987. Extra-pair fertilizations in a predominantly monogamous bird: genetic evidence. Animal Behaviour 35:877–886.

Wheelwright, N. T., K. A. Tice, and C. R. Freeman-Gallant. 2003. Post-fledging parental care in Savannah sparrows: sex, size and survival. Animal Behaviour 65:435–443.

White, G. C., A. B. Franklin, and T. M. Shenk. 2002. Estimating parameters of PVA models from data on marked animals. Pages 169–190 *in* S. R. Beissinger and D. R. McCullough, editors. Population viability analysis. Chicago University Press, Chicago, Illinois.

Whitfield, M. J., K. M. Enos, and S. P. Rowe. 1999. Is brown-headed cowbird trapping effective for managing populations of the endangered southwestern willow flycatcher? Studies in Avian Biology 18:260–266.

Whitlock, M. C. 2002. Selection, load and inbreeding depression in a large metapopulation. Genetics 160:1191–1202.

Whitlock, M. C., P. K. Ingvarsson, and T. Hatfield. 2000. Local drift load and the heterosis of interconnected populations. Heredity 84:452–457.

Willis, E. O. 1972. The behavior of spotted antbirds. Ornithological Monographs 10:1–162.

Wilson, E. O. 1975. Sociobiology, the new synthesis, 2nd ed. The Belknap Press of Harvard University Press, Cambridge, Massachuasetts.

Wilson, P. L., M. C. Towner, and S. L. Vehrencamp. 2000. Survival and song-type sharing in a sedentary subspecies of the song sparrow. Condor 102:355–363.

Wilson, S., and P. Arcese. 2003. El Niño drives timing of breeding but not population growth in the song sparrow (*Melospiza melodia*). Proceedings of the National Academy of Sciences of the United States of America 100:11139–11142.

Wilson, S., and P. Arcese. 2006. Nest depredation, brood parasitism and reproductive variation in a song sparrow metapopulation. Auk 123: in press.

Wingfield, J. C., and K. K. Soma. 2002. Spring and autumn territoriality in song sparrows: same behavior, different mechanisms. American Zoologist 42:11–20.

Wittenberger, J. F., and R. L. Tilson. 1980. The evolution of monogamy: hypotheses and evidence. Annual Review of Ecology and Systematics 11:197–232.

Woodworth, B. L. 1999. Modeling population dynamics of a songbird exposed to parasitism and predation and evaluating management options. Conservation Biology 13:67–76.

Woolfenden, B. E., H. L. Gibbs, S. G. Sealy, and D. G. McMaster. 2003. Host use and fecundity of individual female brown-headed cowbirds. Animal Behaviour 66:95–106.

Woolfenden, G. E., and J. W. Fitzpatrick. 1989. The Florida scrub jay: demography of a cooperative-breeding bird. Princeton University Press, Princeton, New Jersey.

Woolfenden, G. E., and J. W. Fitzpatrick. 1991. Florida scrub jay ecology and conservation. Pages 542–565 in C. M. Perrins, J. D. Lebreton, and G. J. M. Hirons, editors. Bird population studies. Relevance to conservation and management. Oxford University Press, Oxford.

Wright, R. 2004. A short history of progress. Anansi Press, Toronto.

Wright, S. 1922. The effects of inbreeding and crossbreeding on guinea pigs. Decline I. Decline in vigor. Bulletin No. 1090. United States Department of Agriculture, Washington, D.C.

Wright, S. 1969. Evolution and the genetics of populations. Vol. 2: the theory of gene frequencies. University of Chicago Press, Chicago, Illinois.

Wright, S. 1977. Evolution and the genetics of populations. Vol. 3: experimental results and evolutionary deductions. University of Chicago Press, Chicago, Illinois.

Yom-Tov, Y. 1987. The reproductive rates of Australian passerines. Australian Wildlife Research 14:319–330.

Yom-Tov, Y., M. I. Christie, and G. J. Iglesias. 1994. Clutch size in passerines of South America. Condor 96:170–177.

Zanette, L., J. N. M. Smith, H. van Oort, and M. Clinchy. 2003. Synergistic effects of food and predators on annual reproductive success in song sparrows. Proceedings of the Royal Society of London B 270:799–803.

Zirkle, C. 1952. Early ideas on inbreeding and crossbreeding. Pages 1–13 in J. W. Gowen, editor. Heterosis: a record of researches directed toward explaining and utilizing the vigor of hybrids. Iowa State College Press, Ames, Iowa.

Zwickel, F. C. 1992. Blue grouse (*Dendragapus obscurus*). Pages 1–28 in A. F. Poole and F. B. Gill, editors. The birds of North America, number. No. 15. American Ornithologist's Union, Philadelphia, Pennsylvania.

INDEX

Italicized page numbers indicate figures, captions, and tables.